Making & Enjoying
TELESCOPES

Making & Enjoying
TELESCOPES

6 COMPLETE PROJECTS & A STARGAZER'S GUIDE

Robert Miller &
Kenneth Wilson

Sterling Publishing Co., Inc. **New York**
A Sterling/Lark Book

Editor: Holly Boswell
Art Director: Kay Holmes Stafford
Illustrations: Robert Miller & Kenneth Wilson
Photography: Richard Babb, Robert Miller & Kenneth Wilson
Production: Kay Holmes Stafford

Library of Congress Cataloging-in-Publication Data

Miller, Robert, 1948-
    Making & enjoying telescopes : 6 complete projects & a stargazer's
guide / by Robert Miller & Kenneth Wilson.
        p.   cm.
    "A Sterling/Lark book."
    Includes bibliographical references and index.
    Summary: This guide for the backyard astronomer provides basic
information, offers activity ideas, and gives construction details and
assembly drawings for six telescopes.
    ISBN 0-8069-1277-4
    1. Telescopes--Design and construction--Juvenile literature.
2. Astronomy--Observers' manuals--Juvenile literature.
[1. Telescopes. 2. Astronomy--Observers' manuals.] I. Wilson, Kenneth,
1954-   . II. Title.
QB88.M55 1995
522'.2--dc20   95-4592
                                                             CIP
                                                              AC

10 9 8 7 6 5 4 3 2

A Sterling/Lark Book

First paperback edition published in 1997 by
    Sterling Publishing Company, Inc.
    387 Park Avenue South, New York, N.Y. 10016

Produced by Altamont Press, Inc.
    50 College Street, Asheville, NC 28801

© 1995 by Robert D. Miller & Kenneth D. Wilson

Distributed in Canada by Sterling Publishing, % Canadian Manda Group
    One Atlantic Avenue, Suite 105, Toronto, Ontario, Canada M6K 3E7

Distributed in Great Britain and Europe by Cassell PLC
    Wellington House, 125 Strand, London WC2R 0BB, England

Distributed in Australia by Capricorn Link (Australia) Pty Ltd.
    P.O. Box 6651, Baulkham Hills, Business Centre, NSW 2153, Australia

*This book is dedicated to my parents, for their
years of encouragement.*
                                              Robert Miller

To my loving wife, Betty Philpotts Wilson.
                                              Kenneth Wilson

ACKNOWLEDGEMENTS

A debt of gratitude is owed to many individuals whose assistance and encouragement helped make this book possible and to the many telescope makers who have have contributed so much to the art of building telescopes.

A special thanks goes to Dr. Tom Robinson, who loaned us his workshop to get these projects completed in a nearly impossible time frame. Bill Osterhagen contributed a great deal of time, energy and enthusiasm to many of these projects. Tom and Bill are now building their own telescopes. Without the interest and generosity of these two individuals, you would not be reading this book now.

A sincere word of thanks goes to Dr. Allan Saaf, for his contributions to this book and with whom I have had many pleasant hours building more than two dozen telescopes.

A special acknowledgement is due to Mr. Ken Crossen of Techview Corporation for the use of the wonderful Techedit professional illustration software which was used to make the illustrations for these projects.

Both authors would like to thank Rik and Dolores Hill for many valuable discussions on telescope making and observing.

C O N T E N T S

Welcome to the fascinating world of amateur telescope making and to the vast realm of the biggest hobby of all—amateur astronomy!

The telescope is a deceptively simple invention—just a few lenses and possibly a mirror or two. Yet few other pieces of human technology have had such a profound impact on science and philosophy. The telescope is the key that's unlocked the door of the universe to the human mind. Thanks to it , we can now contemplate a cosmos that's billions of light years across and at least 8 billion years old. This marvelous little device kicked our collective egos out of the center of the universe and, at the same time, opened our eyes to marvels stranger than anything we could ever have dreamed.

Almost as amazing as the universe revealed by the telescope, is the fact that anyone can still make one of their own using a few dollars worth of readily available parts. With it, he or she can follow in the footsteps of Galileo or Herschel and recreate their discoveries by observing the same objects they studied hundreds of years ago. And, since many of those heavenly bodies are so far away that it takes hundreds, thousands, and even millions of years for their light to reach our eyes, the telescope is also a time machine which can transport the eye and mind back through the history of the universe! Think of it, for a few dollars and a few hours of your time, you can build your own personal time machine! What's more, even in today's world of high tech and multi-million dollar scientific research, amateur astronomers with modest equipment can still make observations of value to science from their own backyards.

Little wonder that amateur telescope making is

a long established hobby. It grew from the natural human desire to understand the universe first hand, and from the reality that commercially made telescopes have all too often been priced out of reach for the person of average means. In the early 1920's Albert Ingalls, associate editor of Scientific American, discovered articles by architect, artist, and Arctic explorer Russell W. Porter describing how to make a reflecting telescope by grinding your own mirror. This prompted a series of articles on telescope making that inspired a small army of telescope builders across America and elsewhere in the world. Enthusiasm for grinding mirrors and building the telescope to house them has continued unabated ever since.

The most daunting aspect of making a telescope has always been the tedious and error prone grinding of the mirror. Fortunately, today's would-be telescope maker need not grind a mirror to build an economical telescope. Various suppliers offer completed mirrors of high quality for little more than the cost of the materials to grind one. Some amateurs still grind mirrors, both for the personal sense of accomplishment and to create optical designs not commonly available on the retail market. But, the point is, a novice no longer needs to go this route to make an affordable telescope.

This book is designed for the beginner who wants to economically fabricate a telescope and learn how to use it to personally explore the cosmos. If you carefully follow the instructions, you'll wind up with an instrument far better than any of Galileo or Newton's time. With reasonable care, any of the telescopes in this book will provide a lifetime of celestial exploration.

Annular solar eclipse, May 10, 1994. The Moon is grazing the Sun's southern limb, where the ring of sunlight is broken, forming Bailey's beads. This photograph was taken by R. Miller with a Celestron™ 5-inch telescope and a full aperature solar filter.

The Scope of History
A Brief History of Telescopes

Inventor Incognitus

Who made the first telescope? No living person knows the answer. It's one of the great mysteries of science history. We don't even know who invented the first lenses—which may date back to 2000 B.C. or earlier. We do know that spectacles, or eyeglasses, made their first widespread appearance in Europe during the 13th Century. Much later, sometime early in the 17th Century, an unknown dabbler was playing around with pairs of spectacle lenses and managed to hit on the right combination to make distant objects look closer. Thus was born one of the most useful scientific instruments of all time—the telescope.

By 1608, Jan Lippershey, a Dutch spectacle maker, was applying for a patent on his version of the telescope. His application was denied on the grounds that the invention was already widely known by others.

Galileo's Glass Eyes

Within a year, news of this marvelous optical wonder reached Galileo while he visited Venice. As soon as he got home to Padua, he re-created the telescope—his first of many. This first attempt magnified only three times. Later he constructed instruments with lenses up to one and three quarters of an inch in diameter and magnifying as much as 33 times. Galileo's best telescope was very crude by today's standards, even when compared with today's toys.

Nonetheless, Galileo's keen mind, amplified by even a crude telescope, shook the foundations of the astronomical establishment with his celestial observations. The telescope revealed to him vistas of towering mountains and jagged craters on the Moon—the same moon that others fervently believed was a perfectly smooth orb. He saw that the planet Venus waxed and waned through moon-like phases. These phases could only be explained if Venus circled the sun rather than the prevailing belief of it in an Earth centered orbit. The cloudy band of the Milky Way was resolved by the telescope into a cosmic beach composed of thousands of stars not visible as individuals to the naked eye. When Galileo focused his instrument on Jupiter he found the planet had a retinue of four moons circling around it. Saturn baffled Galileo. It seemed to have two ear-like appendages. In later years, to Galileo's astonishment, they seemed to have disappeared.

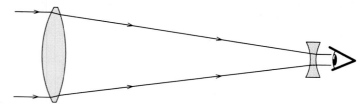

Figure 1: Galilean Telescope

New and Improved?

Galileo's telescopes *(Fig. 1)* used two lenses,

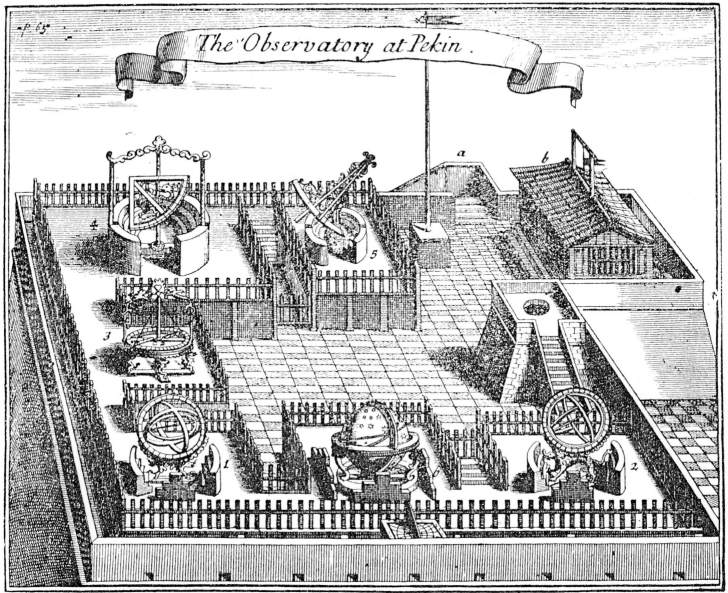

a. Steps going up to the Observatory | 1: a Zodical Sphere | 3 an Azimuthall Horizon | 5 A Sextant
b. A Retiring Room for those that make Observations | 2 an Equinodial Sphere | 4. a Quadrant | 6 a Cælestiall Globe

Ancient observatory at Peiping, China. (from LeConte's Voyage to China, *1698)*

one fatter in the middle than its edges (convex) at the sky end of the tube; and the other thicker at the edges than the middle (concave) located at the eye end. The concave eyepiece gave the telescope a very narrow field of view and greatly limited the magnifying power. The German astronomer Johann Kepler solved this problem by employing a convex eyepiece lens *(Fig. 2).*

As telescopes grew in size and power certain distortions became increasingly evident.

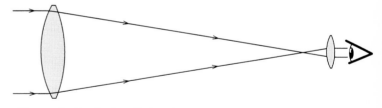

Figure 2: Keplerian Refractor

9

Spherical aberration, one form of distortion, caused the center of the image to have a different point of focus than the outer edge. This made it impossible to get all parts of a large object in sharp focus at the same time. The other major distortion came in the form of chromatic aberration. Chromatic aberration is caused by the inability of single lenses to focus all colors of light at the same point. In other words, the blue light entering a telescope lens from an object—the Moon, for example—would come to focus at a different point than the rays of green, yellow, orange or red moonlight would. Eventually telescope makers learned that they could minimize these distortions by making the main, or objective, lenses of their telescopes with very shallow curves. This led to monstrous telescopes of very long focal length. The Huygens brothers, Christian and Constantine, built one that stretched 210 feet in length! These lanky instruments were very difficult to support and use, but they did give the best quality images of the time. In fact, Christian Huygens was able to resolve Saturn's mysterious 'ears' using a 23 foot long refractor with a lens two and a third inches in diameter. The 'ears' turned out to be the beautiful rings that we now know to circle the planet.

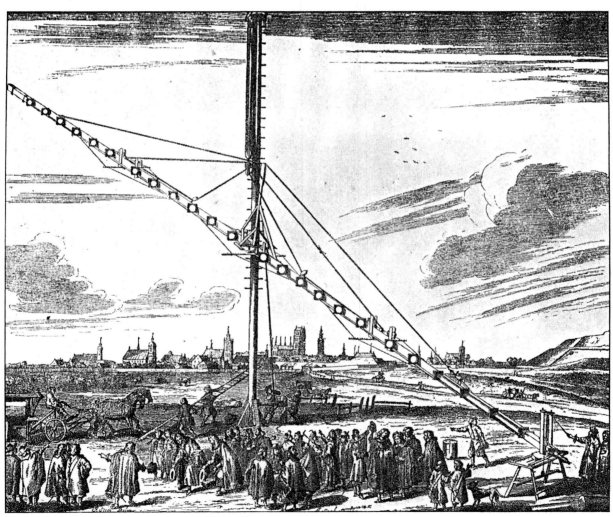

Hevelius' 150-foot telescope being erected, circa 1642.

It's All Done with Mirrors

Other minds searched for ways to make shorter telescopes without the distortions. In 1663 mathematician James Gregory proposed using a concave mirror, rather than a convex lens, to be the objective of a telescope *(Fig.3)*.

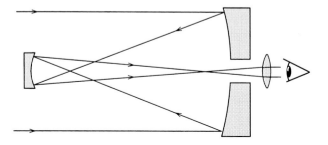

Figure 3: Gregorian Reflector

Rather than being bent to a focus by passing through a lens, Gregory's curved mirror reflected the light to a point of focus. Since the light didn't pass through a glass lens, it wouldn't suffer from chromatic aberration. By making the mirror's shape in the form of a paraboloid, it could bring all points of the image into focus at the same place. This solved the spherical aberration problem. A smaller concave mirror would reflect the light back through a hole in the objective mirror to the eyepiece. The first successful Gregorian telescope wasn't made until 1674. Two years earlier a Frenchman named Guillaume

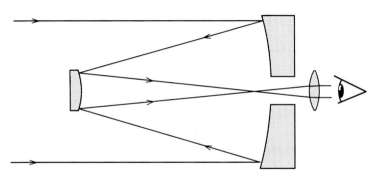

Figure 4: Cassegrain Reflector

Cassegrain proposed a similar design using a spherical primary mirror and a convex secondary *(Fig. 4)*.

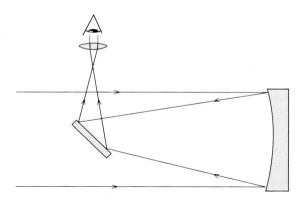

Figure 5: Newtonian Reflector

Perhaps the best known designer of reflecting telescopes was the renowned scientist Isaac Newton. His design used a spherical primary mirror and a flat secondary mirror to bounce the light to the eyepiece on the side of the telescope tube *(Fig. 5)*.

It took a little while for these reflecting tele-

Model of Newton's reflector, circa 1675

scopes to catch on. When they did, the stage was set for a tremendous growth in the size and light gathering power of telescopes.

Bigger is Better

One of the pioneer developers of the reflector was also one of the most famous amateur

Herschel's 40-foot telescope with 48-inch reflector, 1789.

astronomers of all time, William Herschel. A German immigrant to 18th Century England, Herschel was a musician by trade. In his mid thirties Herschel turned his attention back to a boyhood interest in science and math. He purchased a lens and made a small refractor. But he soon found it inadequate for exploring the heavens. In what was perhaps one of the earliest cases of 'aperture fever' Herschel went through an array of larger and more elaborate telescopes, searching for one that would meet his high standards. He soon abandoned refractors and turned his attention to reflectors. When he found he couldn't purchase a reflector large enough to suit him, he decided to make his own. Aided by his sister and brother, Herschel set up a do-it-yourself shop at home and started to make telescopes. There were very few 'how-to' books in those days, so he spent a lot of time on the beginner's slope of the learning curve. Telescope mirrors in those days were made of speculum metal, a mixture of copper and tin. Speculum was notoriously difficult to cast and machine. It's said that Herschel made 200 attempts before he finished his first successful mirror. In 1781 he used one of his early telescopes to discover the planet Uranus—the first time in recorded history that anyone had discovered a new planet. This achievement brought him widespread fame and a lifetime appointment by George III as court astronomer. This freed him from dependence on his musical career and allowed him to pursue his observing and telescope making. Eventually he built the largest reflecting telescope of the day. It had a mirror 48 inches in diameter and a tube almost 40 feet long. It was supported by an elaborate mounting made of scaffolding, ropes and pulleys.

The apex of giant speculum reflectors was erected in 1845 on the island of Ireland. This *Leviathan of Parsonstown,* as it was known, had a four ton mirror with a diameter of 6 feet. This behemoth was mounted between two massive masonry walls, each 56 feet high and 72 feet long. This record breaking telescope was the brain child of Irishman William Parsons, the Earl of Rosse. Unfortunately, the cloudy and damp Irish climate was less than ideal for productive use of telescopes, especially those with speculum mirrors.

Off with the Rose Colored Glasses!

As the speculum eyes of the reflectors widened by an order of magnitude, refracting lenses took a major step forward in quality, if not in size. London opticians in the middle of the eighteenth century learned to correct most of the chromatic aberration that had plagued refractors since starlight had first entered a tele-

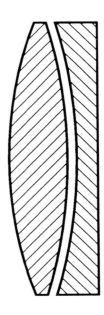

Figure 6: Achromatic objective

scope. They discovered that, by adding a secondary convex lens *(Figure 6)* made of flint glass to the single convex crown lens, the chromatic aberration of one lens would cancel the chromatic aberration of the second lens. Even today most refractors use such two piece *achromats*, as they're called, for objectives. This invention greatly improved the quality of refractors. But, unfortunately, the glass making technology of the day was very primitive. No one seemed to be able to make optical quality flint glass disks any bigger in diameter than about four inches. This roadblock assured the dominance of mirror telescopes in the medium to large size aperture until well into the 1800's. Eventually glass making improved and refractors up to 40 inches in diameter were built. Unfortunately, the lenses of refractors reach a practical limit at about 40 inches. Larger lenses will sag from their own weight and ruin the precise shapes that they need to form good images.

Repairing Tarnished Reputations

Unlike lenses, which can only be supported around the edges, mirrors can be bolstered from behind. Although less limited in size, speculum mirrors had other problems. One of them was that they tarnished very easily. This resulted in a great loss of precious light and forced Herschel and his contemporaries to frequently re-polish their mirrors. This became especially tedious with the larger mirrors that weighed up to a ton or more. Even when freshly polished, speculum only reflected about 40% of the light striking it.

An important development in the 1850's was to brighten the tarnished reputation of reflecting telescopes. A method of coating glass with a thin layer of silver was developed. By 1856 Frenchman Leon Foucault and Carl August von Steinheil of Munich applied this technique to telescope mirrors. These silver mirrors turned out to be about 50% more reflective than speculum. And, just as important, the glass was lighter in weight and easier to grind. The silver coatings still had a habit of tarnishing, but re-coating them was nonetheless easier than re-polishing speculum.

Silver Eyed Monsters

The fundamental optical design of astronomical telescopes hasn't changed greatly during the twentieth century. Lenses or mirrors are still used to funnel light to a focus. Limited to a maximum of 40 inches in diameter, refractors have not been able to keep up with their shiny

Nasmyth's 20-inch Cassegrain-Newtonian, circa 1845.

siblings, the reflectors, at least when it comes to blazing the distant frontiers of modern research astronomy. That territory demands every photon of light that can be gathered and requires very large aperture.

Refractors are still with us in the form of binoculars and telescopes used by amateur astronomers. Indeed many of the finest quality telescopes in the 3 to 8 inch diameter range today are refractors. Many of them employ new high tech glasses and optical designs unheard of a hundred years ago.

Today's reflectors use low expansion glasses like Pyrex™ and other ceramic glasses usually coated with aluminum rather than silver. Aluminum is almost as reflective as silver yet lasts about ten times longer before it tarnishes to the point of needing re-coating.

Several hybrid telescope designs have become quite popular in certain niches. *Catadioptric* telescopes combine the concave mirror of the reflector with a thin correction lens. These designs further improve the image qualities of the reflector and allow for wider fields of view. Two notable designs in this category are the Maksutov and the Schmidt-Cassegrain. These are often found as commercially made telescopes in the 3 to 14 inch range.

The giant reflectors of today's professional astronomers work much as Isaac Newton's did. But, in most cases, the human eye has been replaced by much more sensitive and objective devices—photographic plates and CCD electronic sensors. These can be placed right at the focal point of the giant mirrors without the need of a secondary mirror to divert the light out to the side of the tube. Secondary mirrors are often used, however, to extend the focal length of the primary mirror for certain kinds of research. Thus most giant reflectors have

holes in their primary mirrors to allow them to be converted to Cassegrains as needed.

Although much easier to make in large sizes than refractors, there are practical limits on the size of the mirrors of reflectors. Conventional reflectors topped out with the 236 inch mirror of the Bolshoi Alt-azimuth Telescope in the former Soviet Union. Old technology mirrors much larger than around 200 inches become engineering nightmares.

In the late 1970's new technologies and designs surfaced that, while not quite breaking the 200 inch barrier, allowed astronomers to sneak around it. The first of these involves taking the light gathering power of several smaller mirrors and combining them to form a single image. The prototype for such a telescope is the Smithsonian Astrophysical Observatory's Multiple Mirror Telescope. This instrument combines the light of six 72 inch mirrors to produce the equivalent of a telescope with a single 176 inch mirror. Sophisticated computers and laser alignment systems are needed to maintain the critical alignments of such a system. The results from this initial effort were so successful that many larger multiple mirror telescopes have been designed and constructed. Many more are on the drawing boards. One of grandest members of this new generation of telescopes is the recently completed Keck telescope located atop Mauna Kea in Hawaii. It uses 36 hexagonal mirrors, each one about 72 inches across. Their combined light gathering power equals that of a single mirror 400 inches across!

CHAPTER 2

But Where Do I Look?
The Inside Story of How Telescopes Work

Light Work

Light is a curious commodity. Physicists have decided to treat the stuff as if it were two different beasts. On one hand it behaves as if it were made of extremely small billiard balls which zip off at some 186 thousand miles per second (300,000 km/sec), ricocheting off any solid, opaque surface they hit. On the other hand, light emanates in waves, like ripples on the surface of a pond. Depending on the situation, either of these models can be very useful in describing light's behavior. Both are helpful in understanding how telescopes work.

Let's start with particle light and mirrors. When a beam of light hits a mirror, it bounces off. If that beam of light hits the mirror perpendicular to its surface *(Figure 7)* it will reflect right back the way it came. That's why you always see yourself when you stare straight into the bath-room mirror. If the rays of light strike the mirror at an angle, the light will bounce back at the same angle *(Figure 8)* as it hit. This property of flat mirrors permits them to be used in everything from submarine periscopes to your car's rear view mirror.

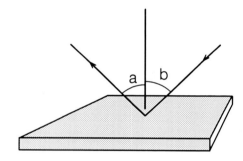

Figure 8: Reflection Angle = Incident Angle

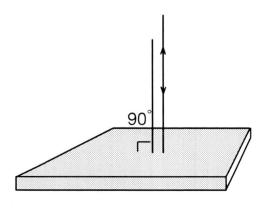

Figure 7: Reflection from a flat mirror

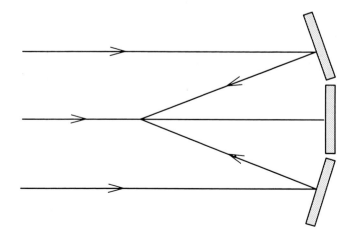

Figure 9: Light converged with three flat mirrors

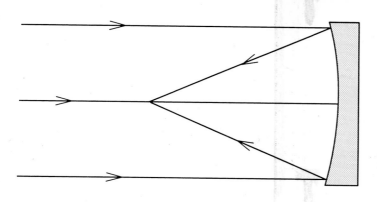

Figure 10: Reflections from a concave mirror

Waves passing from air into a flat piece of glass *(Figure 11)*, for example, slow down momentarily. Even though they slow down, they remain parallel and moving in the same direction. When they reach the other side of the glass sheet and return to air, they speed up again and maintain their original direction.

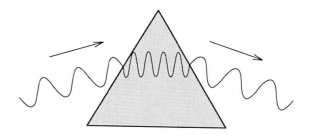

Figure 12: Ray of light passing through a prism

If we arrange a series of small flat mirrors, each set at the proper angle *(Figure 9)*, we can use them to reflect light coming from one direction to a single point called the focal point. Notice that each mirror marks a point along a concave curve. If we replace this series of mirrors with a single mirror *(Figure 10)* whose surface is curved in shape to match that of the arrangement of small mirrors, it will concentrate the incoming light in the same way, only more effectively. All reflecting telescopes gather light and concentrate it to a focus in this way.

The objective lens of a refractor reaches the same goal by different means. If we think of light as a series of parallel waves we can look at what happens when these waves pass from one medium into another.

If the waves instead pass into a triangular piece of glass called a prism, their direction of travel is changed. As a wave hits the side of the prism, one end of the wave hits first and slows down. The other end of the wave travels faster a bit longer before it hits the glass. This causes the wave to change direction *(Figure 12)*. This bending of light is called refraction.

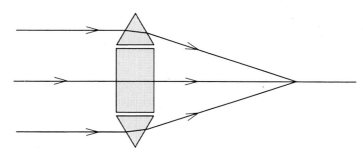

Figure 13: Light converged with two prisms and a cube

By placing two identical prisms on either side of a similar size cube of glass *(Figure 13)* we create a device which takes three parallel beams of light and bends them together to a focus. If these three pieces of glass were combined into

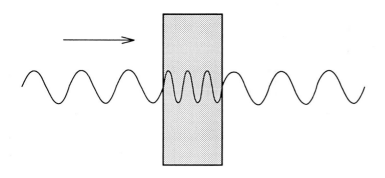

Figure 11: Light waves passing through flat glass

one and the angles smoothed into a curve *(Figure 14)*, we'd have a simple convex lens.

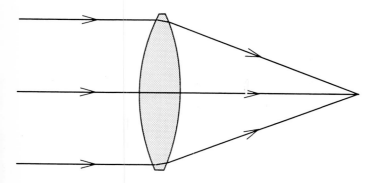

Figure 14: Light passing through a convex lens

The distance from the telescope objective, be it lens or mirror, to its focal point is called the focal length of the objective. The focal length of the objective divided by the diameter of the objective gives us a number called the focal ratio, or f-ratio. A 10 inch (25.4cm) mirror, for example, with a focal length of 80 inches (303.2cm) has an f-ratio of 8, commonly written as f/8.

No matter how the light is concentrated, a telescope has three prime functions: to gather light, to show fine detail, and to magnify the image.

Catching Some Rays—Light Gathering Power

In order for us to see a distant object better, the telescope's objective has got to gather more light than the human eye alone can and funnel it to a focus where an image is formed. The more light the objective gathers the brighter that image will be. Similarly, the more light gathered, the fainter the object that can be seen with the telescope. The light gathering power of the objective depends on two factors: the clear area of the lens or mirror, and the efficiency of that

lens or mirror. All lenses scatter and absorb a certain amount of light falling on them, and no mirror is a perfect reflector. From 35% to 40% of the incoming light may be lost in a typical telescope by the time it gets to your eye. This is usually more than made up for by the much greater surface area of the telescope objective compared to the naked eye. At its widest, the opening of the human eye is about 0.3 inches (0.76cm) in diameter. A very small refractor with an objective lens one inch in diameter has a surface area 11 times greater than the naked eye. A 6 inch (15cm) diameter telescope objective gathers over 470 times as much light as the naked eye! Bottom line—the greater the objectives diameter, the greater the light gathering power of the telescope.

Reading the Fine Print—Telescope Resolution

The diameter of the objective also determines the telescope's resolution, or ability to see fine details. This is usually described by the angular separation of the two closest stars that can be seen as separated by the telescope. The better the resolution, the closer together a pair of stars can be and still be seen as separate stars. The separations are typically measured in seconds of arc, where a single degree of angle is subdivided into 60 equal minutes of arc and each of those minutes is divided into 60 equal seconds. Thus one second of arc equals 1/3600 of a degree.

English amateur astronomer William Dawes worked out a rule of thumb for what the ideal resolution of a telescope should be. In seconds of arc it equals 4.56 divided by the diameter of the lens or mirror in inches. Thus a 4.5 inch (11.43 cm) telescope should resolve a double star separated by one second of arc. A nine inch should split a pair of stars half an arc second apart, and so on. Actual resolution also depends on the quality of the telescope's optics, their alignment, the number and size of any obstructions in the light path and the steadiness of the earth's atmosphere. Even at the best of times, the earth's atmosphere seldom is steady enough to allow resolution better than one arc second, no matter how big the telescope.

Power Mad—Telescope Magnification

The objective lens or mirror creates a small bright image of the distant object. Once this image is formed we can use a smaller lens or combination of lenses called the eyepiece to magnify that image. An eyepiece is, in effect, a small microscope. *Magnifying power is the least important of the main telescope features, and the biggest pitfall for gullible and unknowledgeable telescope buyers*.

The amount of magnification depends on the focal lengths of the main mirror or lens and that of the eyepiece. To find the magnification, or power, of a telescope divide the focal length of the telescope's objective by the focal length of the eyepiece it uses. For example, a telescope with a mirror having a focal length of 50 inches and an eyepiece 1 inch in focal length magnifies 50/1, or 50 times (usually written 50x). Likewise, a refractor with 1000 millimeter focal length and a 10 millimeter focal length eyepiece will magnify 1000/10, or 100 times. Notice that you must be consistent in units when you calculate magnification. If the focal length of the telescope's objective is described in inches, you must use the eyepiece's focal length in inches too. To help you convert from inches to millimeters, keep in mind that one inch equals 25.4 millimeters.

Most telescopes allow you to use different eyepieces. By inserting eyepieces of different focal lengths, you can vary the magnifying power of the telescope. In theory, you can make any telescope magnify any amount you want, just by choosing an eyepiece of the right focal length. There is, however, a practical limit. This limit is determined by the amount of light that's been gathered by the objective, which is dictated by the collecting area of the objective. A simple rule of thumb is that the practical magnifying limit is about 50 times the diameter of the objective in inches, or 2 times the objective diameter measured in millimeters. For example, a refractor with a lens two inches in diameter has an upper power limit of about 100x. Any greater power, while technically possible, gains nothing. Furthermore the 50x per inch guide reaches a practical limit at telescopes of 10 to 12 inches in diameter. So, no matter how large the objective, the atmosphere is rarely, if ever, steady enough to allow powers of more than 500x to 600x.

Higher powers also come with drawbacks. The more you magnify something in a telescope, the more you spread out its light. Remember the objective gathers only a fixed amount of light. The more you magnify, the more thinly you spread out that light. Eventually the image gets so dim you can hardly see it anymore. When you magnify the image, you also magnify any unsteadiness in the air and any shakiness in the supporting stand of the telescope. Under high power, the slightest breeze can shake the telescope to the point where the image looks like Jello™ in an earthquake! In addition, generally speaking, the higher the power, the smaller the area of the sky (field of view) you get to see. For large objects that means you might only get to see part of the object at one time. It also makes searching for an object much harder.

The majority of experienced amateur astronomers use their lowest power eyepieces most often, especially for finding objects and looking at larger and fainter targets. They save the higher powers for planets, double stars and small objects. Even then, they shy away from the highest powers unless the air is very steady.

But Where Do I Look? Telescope Types

Perhaps the most popular amateur-built telescopes are the Newtonian reflectors. Usually they provide the most light gathering power for the money. Just as with Sir Isaac's first one, these telescopes have a concave mirror located at the bottom of an open tube. This mirror bounces the light back up the tube to a focus. Before the cone of light reaches the focal point, a smaller, flat mirror, called a diagonal, diverts the light out through a hole on the side of the telescope tube and on to the eyepiece. The diagonal and its support structure obscure a small amount of the incoming light, but usually not enough to be a great problem. And, in case you're worried, they are not visible when you look through the telescope at a distant object. Due to the flat secondary mirrors, all Newtonian telescopes have the observer facing at a 90 degree angle to the direction the telescope is pointed.

In years gone by, many amateurs ground their own mirrors to save on construction costs. For common sizes and focal lengths it's not too difficult to grind a good mirror. All you need is some spare time and the ability to carefully follow directions. On the other hand, it is time consuming and can be messy. Nowadays, you can buy a professionally made mirror for not much more than the cost of the materials to grind one. The result is that fewer and fewer hobbyists make their own mirrors anymore. To make it as simple as possible for the beginner, the projects in this book will use finished optics, so you won't have to worry about telescope construction becoming a grind!

To most people the classic telescope design (at least for pirates and mad scientists) is that of the refractor *(Figure 15)*. Here an objective lens, usually with two or more elements, gathers the light and sends it one way down the closed telescope tube to the eyepiece at the opposite end. When using a refractor the observer may look straight through the telescope or, if the angle of the object is awkwardly high, insert a device called a star diagonal. The star diagonal contains a small prism or flat mirror which redirects

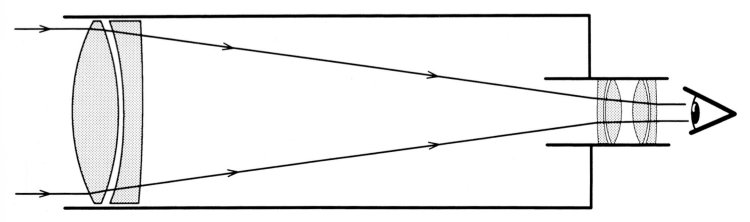

Figure 15: Refractor

the light at a right angle to the telescope tube. A few amateurs grind their own objective lenses, but this is rare. As with reflectors, it's more common to purchase a quality refractor objective and build a telescope around it.

Both Newtonian reflectors and classical refractors have tubes almost as long as the focal lengths of their objectives. This usually isn't a problem when the focal length is under about 5 feet (150cm). Longer focal lengths require tubes that quickly get unmanageable, especially if you need to haul the telescope outside every time you use it or pack it into a compact car for transportation to a good observing site. A common solution to this problem is to fold the optical path of the telescope. The Gregorian and Cassegrainian compound designs were early examples of this. Even more popular today, among commercially made telescopes, are the catadioptric designs. They combine a compound reflecting system of two mirrors with a thin correcting lens. These include the Maksutovs and Schmidt-Cassegrains. These telescopes are very portable for their long focal lengths, but complex enough to place them outside the realm of all but the most advanced telescope maker.

Supporting Acts

The design and optics of a telescope are but one part of the telescope system. In order to study the magnified image formed in the telescope, the optical tube must be firmly supported by a structure called a mounting. Ideally this mounting should allow you to easily point the telescope anywhere you'd like to in the sky and to track the object you want to study.

Telescope mountings must be strong enough to support the optical tube and rigid enough that any vibrations generated by the wind or

your touching of the telescope will dampen out quickly. Remember that the magnification of the telescope will also magnify any vibrations of the mounting by a proportionate amount. The mounting must also balance the weight of the tube assembly so that it will move easily but without drifting due to gravity.

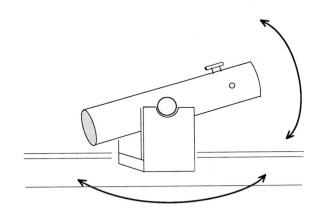

Figure 16: Telescope in alt-azimuth mount

Telescope mounts come in two fundamental designs: *alt-azimuth* and *equatorial*. Alt-azimuth mountings *(Figure 16)* are the simplest and least expensive ones. They work like an artillery gun or a surveyor' s transit. One axis of rotation is perpendicular to the ground. It allows you to point the telescope in any direction along the horizon. This is the azimuth motion. 'Alt' stands for 'altitude', or the angle of the telescope up from the horizon. By moving the telescope in a combination of these two motions, you can aim it anywhere on the hemisphere of sky overhead. A very popular version of the alt-azimuth design was developed by John Dobson of the San Francisco Sidewalk Astronomers. The Dobsonian mount uses relatively inexpensive materials (plywood, Formica™, and Teflon™) to make a compact and sturdy telescope support favored for portable, large aperture Newtonians.

A Tough Act to Follow

Astronomical objects have an annoying compulsion to slowly move across the sky, or at least seem to. What's really happening is that Earth-bound telescopes sit on a moving platform. As the Earth does a daily pirouette round on its axis it gradually changes the direction any telescope points in space. This causes heavenly bodies to drift out of the field of view of the telescope. The higher the telescope power, the faster objects drift out of view. That is, unless the telescope slowly moves in precisely the opposite direction at the same speed.

By frequently turning the telescope by small amounts you can keep it aimed at a sky object. The higher the magnifying power, the more often you'll have to re-aim the telescope. Unfortunately, unless you live on the equator or at the north or south pole, celestial objects don't move in a simple azimuth only, or altitude only direction. In most cases an alt-azimuth mounting will need to be adjusted in both axes to track a celestial object. This can become very tedious, especially at higher powers.

Long ago this problem led astronomers to design equatorial mountings. One axis of an

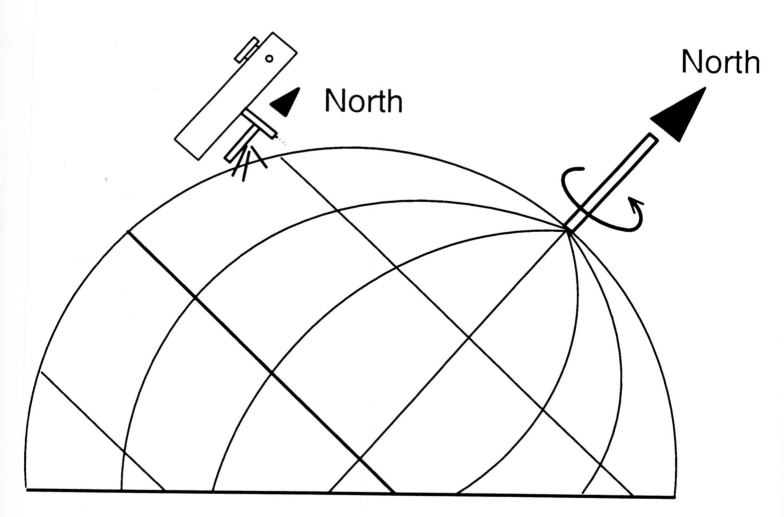

Figure 17: Telescope polar axis parallel to Earth's rotation axis

equatorial mount is designed to be oriented parallel to the Earth's axis of rotation *(Figure 17)*. This is usually called the *polar axis*. By turning the telescope around this one axis, you can compensate for the Earth's rotation. Since the Earth's rotation rate is practically constant, a *clock drive* (a motor and set of gears) can be attached to the polar axis to make the tracking almost automatic. This frees the observer to spend time enjoying the view, rather than constantly re-aiming the telescope.

There are many design variations of the equatorial mount. Perhaps the two most common are the *German equatorial mount* and the *fork mount (Figure 18)*.

Which is better, equatorial or alt-azimuth? Alt-azimuth is usually least expensive and less complicated to build. It's ideal for observing when you won't be using very high powers often and where you won't be staring at the same object for hours on end.

If, on the other hand, you want to take long exposure photographs or electronic images, or you plan to spend hours looking at small craters on the Moon and fine planetary features, then an equatorial mount is what you want. In building your own telescope you can start with an alt-azimuth mount for simplicity and economy. Then, after you've had some experience using it, you can transplant the telescope to an equatorial mount if you wish.

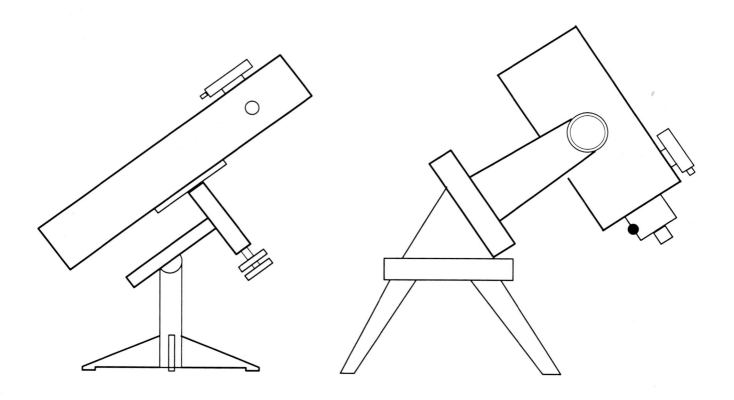

Figure 18: German equatorial mount and fork mount

CHAPTER 3

Bells & Whistles
Telescope Accessories

There are many ways to augment a telescope. Some of them vastly improve a telescope's ease of use. Others are nice black holes to pour money into.

Ready, Aim, Finder!

Perhaps the most useful accessory that any telescope can have is a good *finder*. These are small, low power refractors with wide fields of view and a cross hair visible through the eyepiece. They act as a sort of telescopic gun sight. When properly aligned with the main telescope, they allow you to more easily find objects. With a much wider field of view (typically around 5 degrees) the finder has a better chance of picking up an object after you've first aimed the telescope in the general direction of your celestial quarry. Even the lowest power eyepiece in the main telescope may not have a wide enough field of view to easily find an object. Once you can see the object in the finder, you just carefully move the telescope until the object is centered in the finder cross hairs. Now it will be visible in a low power eyepiece of the main telescope.

Finders don't need to have high magnifying power, 5X to 8X is typical. What is important is that the objective lens of a finder be as large as practical. The bigger the finder objective is, the fainter the object it can see. Often you may be able to spot even faint objects right in the finder.

If not, you will at least be able to see more faint stars which will make it easier to locate your targets. Any finder is better than none. But it's best to have one with at least a 35mm objective. A 50mm objective is much better and some amateurs have finders with 70mm lenses.

Some finders have star diagonals to make them easier to use at awkward angles. Most star diagonals, however, give a mirror reversed image of the sky which can be quite confusing when trying to find something using a star chart. Some observers also feel that it's easier to visualize where the finder is pointed by looking straight through it with one eye while keeping the other eye open. This provides a pseudo-stereoscopic effect that can be quite helpful. For these reasons, many amateur astronomers prefer a straight finder without a star diagonal.

When observing from urban or suburban areas, there's usually enough ambient light from civilization to give the sky seen in the finder a dull gray appearance. This back lighting allows you to see the finder cross hairs without much difficulty. When observing from a location where the sky is very dark you might have a little trouble seeing the cross hairs. For this eventuality, some finders have a built-in light emitting diode or small light bulb to illuminate the cross hairs.

Old flea market binoculars can make good, inexpensive finders. If one half of the binoculars is undamaged, separate it from its twin and devise a mount for it on your telescope.

In recent years a new type of finder has become quite popular. It has no magnification and no lenses. Using flat pieces of glass it allows you to look through a small window and see a naked eye view of the sky with a red cross hair reticle, seemingly projected on the sky. Such finders give you the widest possible field of view and are ideal for aiming a telescope at bright objects.

If your budget is very tight, you can mount a pair of eye screws on the tube a foot or more apart to serve as an aiming device. For a small rich field telescope this may be all that you need.

Power Observing - Telescope Eyepieces

After the telescope objective, the eyepiece is the most critical component in determining the quality of the final image. Get the best eyepieces you can afford. With proper care, they'll last a lifetime. If you upgrade to a better telescope, good eyepieces can travel with you and save a lot of start up costs of a new telescope.

There are several vital statistics about eyepieces that are important to know. We've already mentioned focal length, which determines the magnification that an eyepiece will give in a telescope of a given focal length. The design of an eyepiece, together with its focal length, determine its field of view and *eye relief*. Eye relief is how far back, or how close, your eye must be to see the entire field of view created by the eyepiece. Some eyepiece designs have more eye relief than others for the same given focal length. Within the same eyepiece design, the shorter the focal length (i.e. the higher the power) the less eye relief. This aspect of an eyepiece can be especially important if you have to wear eye glasses to look through the telescope.

Near and far sighted observers can usually take their glasses off and just refocus the telescope. If, on the other hand, you have astigmatism you'll probably have to keep them on for low magnifications. Eyeglasses at the eyepiece naturally force you to keep your eye back a minimal distance from the eyepiece. If that distance is greater than the eye relief of the eyepiece you will lose some of the field of view. This is usually not a great problem with lower power, longer focal length eyepieces but it can become a major headache with shorter focal lengths. If your astigmatism isn't too severe, try viewing with shorter focal length eyepieces without your glasses. Often the smaller cone of light from higher power eyepieces will provide a good image if the astigmatism is mild.

At the other end of the power range we need to be concerned about the *exit pupil* of the telescope-eyepiece combination. The exit pupil is the diameter of the cone of light that leaves the eyepiece and enters your eye. If the exit pupil is wider than the pupil of your eye, much of the light gathered by your objective will be wasted. At its very best the pupil of the human eye will dilate to about 8 mm in diameter. A more typical value for the fully dark adapted, young eye is about 7mm. Older eyes and those not fully dark adapted (which is common around the city lights) seldom get wider than 5 or 6mm. Any exit pupil bigger than these values wastes light. Numerically, the exit pupil's diameter is equal to the diameter of the objective divided by the magnification. For example, a 6 inch diameter reflector with a 36 inch focal length and a 1 inch focal length eyepiece magnifies 36x. The exit pupil of this combination is 6/36, or 0.17 inches. This converts to a 4.2mm exit pupil, well within a 5mm limit. A 2 inch eyepiece, on the other hand, in the same telescope gives a mag-

Assorted eyepieces on a star chart.

nification of 18x and an exit pupil of 8.4mm. This is bigger than the largest dark adapted human pupil and thus would waste light.

Eyepieces come in three standard sizes, all determined by the outside diameter of the part of the eyepiece which slides into the eyepiece holder of the telescope. Lower priced telescopes imported from Japan usually have a standard barrel diameter of 0.965 inch (2.4511cm). The American standard is 1 1/4 inches (3.175cm). In recent years, 2 inch (5.08cm) diameter eyepieces have become very common for long focal length, low power eyepieces. While it's possible to adapt smaller diameter eyepieces to fit the larger eyepiece holders, it's more difficult and less useful to try and fit a larger eyepiece in a smaller holder. The greatest variety of quality eyepieces currently available are found in the 1 1/4 inch size.

Better eyepieces have several things in common. Coated lenses are one of them. The coatings, usually of magnesium fluoride, reduce lens flare and ghost images while also improving the amount of light transmitted through the eyepiece. Single coated lenses usually reflect a bluish tint when held at certain angles under bright white light. Even more desirable are multi-coated lenses. These have a purple or greenish tint. Another feature of a good, modern eyepiece is having a barrel that's threaded for filters.

There are many eyepiece designs on the market today. They range from simple and relatively inexpensive to massive multi-element types that cost more than a good telescope mirror. Here's a summary of the most common designs:

Huygenian. This is the original two lens eyepiece design and the least desirable. They have apparent fields of view of 25°-40° and eye relief of about 25% of the eyepiece focal length. They can be adequate for telescopes with long focal lengths but in any telescope faster (less) than f/8, image quality goes out the window. These eyepieces are the least expensive to make and so are still found on the cheap, imported, 'department store' telescopes.

Ramsden. A primitive design, not much better than the Huygenian. It also uses two lenses to yield apparent fields of view of 30°-40° and eye relief of around 30% of the eyepiece focal length. Again, these eyepieces find their best performance in telescopes with f-ratios slower (greater) than f/8.

Kellner. The Kellner design has three lenses and gives much better images than the Ramsdens or Huygenians. Here apparent fields of view are 35°-45° and eye relief can be 30-80% of the focal length of the eyepiece. These eyepieces perform well with telescopes as fast as f/6. If you're on a tight budget, Kellner's are where you want to start shopping.

RKE™. These are an improved Kellner design which provide wider fields of view and better corrected images at a bargain price . Apparent fields of view are 45° and eye relief is around 87% of the focal length.

Erfle. Erfles are known for their wide apparent fields of view, typically 50°-70°. Eye relief is between 30% to 40% of the focal length of the eyepiece. The Erfle design works well with telescopes as fast as f/4.5. Although its six element design often produces annoying image ghosts when used on planets at higher powers, it's a wonderful eyepiece for wide-angle, low power observing of deep sky objects like galaxies and nebulae. This design has been improved on in recent years by various manufacturers. These new and improved wide field eyepieces will be found under the names *Konig™*, *Tele Vue Wide-Field™*, and *Meade Super Wide Angle™*.

Orthoscopic. In the 1960's and 1970's these were the top of the line eyepieces, especially for medium and higher powers. Orthoscopics have four lenses and apparent fields of 40° to 45°. The eye relief is very generous at about 70-100% of the eyepiece focal length. Excellent performance is had in telescopes with f-ratios as fast as f/4.5. These eyepieces are especially good for higher power views of the Moon, planets and double stars.

Plossl. One of the most popular designs on the market today, the Plossl has four lens elements and produces apparent fields of around 50°. Eye relief with this design is excellent—usually at 70-80% of eyepiece focal length. Although its eye relief is slightly less than the orthoscopic design, many amateurs favor it because of its wider apparent fields of view.

Nagler™ and **Meade™ Ultra Wide**. These new designs give the ultimate in wide field observing. The field of view is over 80° with sharp stars images to the edge of that wide field. Eye relief is usually over 100% of focal length, sometimes over 150%. The longer focal length eyepieces of these designs are so massive that they may require rebalancing your telescope to use them. The only other drawback is the price—over three times what you'd pay for an orthoscopic or Plossl.

Barlow. Strictly speaking, a Barlow is not an eyepiece in and of itself. It's a lens, or system of lenses, with a negative focal length that's inserted into the eyepiece holder in front of the eyepiece. This negative lens effectively lengthens the focal length of the telescope objective by 1.5x to 4x. This increases the magnification of the eyepiece by the same amount. There are two main advantages to using a Barlow. First they allow you to increase the power of a given telescope without switching to a shorter focal length eyepiece. This is helpful if you must wear eyeglasses at the telescope because the Barlow, while increasing the power, doesn't decrease the eye relief. Second, if you're on a budget, and can only afford one or two eyepieces, a well chosen Barlow can effectively double the number of magnifications available to you. For example, a set including 18mm and 7mm eyepieces and a 1.5x Barlow would allow you 56x, 143x, 83x, and 214x in a telescope with a focal length of 1000mm. Some observers feel that Barlows make the images a bit softer than they would be by simply using an eyepiece of the equivalent focal length. If you decide to invest in a Barlow, get one of the better ones. Bargain basement Barlows aren't worth bothering with.

Zoom. Ocassionally you may find zoom eyepieces advertised for telescopes. In order to provide a continuous range of focal lengths, these eyepieces have to make many compromises and add additional lenses. The result is loss of light, loss of field of view, and image ghosts. Save your money and invest in a set of the best fixed focal length eyepieces you can afford.

Eyepiece Recommendations. Your first eyepiece should be the one you'll use the most. Experience has shown that most observers use their lowest power eyepiece the most. So choose the best long focal length eyepiece you can afford—at least a Kellner, or better yet, an Erfle. Pick a focal length somewhere between 20mm and 40mm. Next acquire either a Barlow or a medium focal length Plossl or orthoscopic between 10mm and 15mm. Lastly, get a high power orthoscopic or Plossl with a focal length of around 6mm or 7mm. If price is no object, consider a Konig™, Nagler™ or Meade Ultra Wide™, especially around 13mm as a low to medium power eyepiece.

If you're really pressed for money and good at scrounging and adapting, there is an alternative source for eyepieces. Many of the lenses that are found on old movie cameras and projectors, especially the 16mm movie format, can be used as eyepieces. You can sometimes find these for five or ten dollars at flea markets and yard sales. The camera or projector doesn't have to work since all you want are the lenses. If the lens barrel isn't the right size to fit your eyepiece focuser, you'll have to adapt it. If it' s only slightly smaller, you can wrap it with tape until it fits. Larger gaps can be bridged with cylindrical collars made of metal, wood, or plastic.

Color by Numbers - Telescope Filters

Astronomers often use filters on their eyepieces, especially to bring out subtle planetary detail; reduce the polluting effects of artificial lighting; or to increase the contrast of certain nebulae. These filters usually thread into the telescope end of an eyepiece. There is a standard size thread for 1 1/4 inch eyepieces used by almost all filter and eyepiece manufacturers, except for Brandon.

Among the most useful color filters are light red (No. 23A) which is good for bringing out detail on Mars; green (No. 58) for polar details on Jupiter and Mars; and, blue (No. 80A) to bring out cloud details on Jupiter and Mars. Polarizing and neutral density filters are also handy for reducing glare from bright objects like the Moon and Venus.

Modern technology has produced special filters that can tune out certain wavelengths of light that are common in the artificial outdoor lighting that is becoming an epidemic around our cities and suburbs. This 'light pollution', as astronomers call it, threatens to hide even the brightest stars from future generations. Even now, it makes certain objects very difficult or even impossible to see with a telescope located near even small cities. Light Pollution Reduction, or LPR, filters are designed to cancel out certain wavelengths of light given off by the artificial lighting. These filters can be useful for deep sky observing around a city, especially if most of the light pollution comes from mercury vapor or low pressure sodium lights. These light sources give off most of their light in very narrow wavelength bands which are easier to filter out. Unfortunately, light pollution comes increasingly from high pressure sodium lights which emit over a broad range of the spectrum. Any filter that effectively blocks high pressure sodium will also block most of the visible light that you want to see from a distant galaxy or nebula. Nonetheless, if your observing site is seriously impacted by city lights and you want to view faint objects, you might want to experiment with LPR filters.

A variation of the LPR technology led to special filters designed to block most light except specific wavelengths which certain celestial objects give off in abundance. Planetary nebulas, for example, emit most of their light from doubly ionized oxygen atoms. A filter blocking most light except this oxygen (OIII) emission are a great help in viewing these objects. Similarly, certain diffuse nebulae shine mostly in hydrogen beta light which can be enhanced by using an H-beta filter. These filters tend to be expensive but, though not a panacea, can be useful in certain situations.

Dial-a-Star - Telescope Setting Circles

Examine any globe of the Earth closely and you'll find that it's probably covered with a network of fine markings known as lines of latitude and longitude. These coordinates allow us to locate cities and other points on the earth's surface. The celestial globe has a similar system of coordinates called *right ascension* and *declination*. Declination (dec.) is the heavenly equivalent of latitude. Its lines run parallel to the celestial equator (the projection of earth's equator on to the sky) which marks the $0°$ line of declination. Declination increases as you move away from the celestial equator until it reaches a value of $90°$ at the north and south celestial poles. Declination is considered negative (-) if it's south of the celestial equator and positive (+) when it's north of the celestial equator. The counterpart to earthly longitude is right ascension. Right ascension (R.A.) lines run from the north celestial pole to the south celestial pole at right angles to the celestial equator. Instead of angular degrees, however, right ascension is measured in hours and minutes. There are 24 hours of right ascension, each made of 60 minutes. Sirius, the dog star, for example, is located at celestial coordinates of 6h45m in right ascension and $-17°$ declination.

Many equatorial telescopes have calibrated disks on each axis which allow you to use these celestial coordinates to locate unfamiliar or faint objects in the sky. These disks are called *setting circles* and they can be very helpful, especially when you don 't know your way around the sky very well. Setting circles come in a range of sizes. Usually the bigger in diameter they are, the more precise they can be. Accurate use of setting circles requires that the telescope's polar axis be accurately aligned with the Earth's axis.

The computer revolution has also reached the technology of telescope setting circles. For a price, you can now obtain digital setting circles for you telescope. Some of these are even designed to work on alt-azimuth mounts. The latest of these devices have built-in memories loaded with the locations of thousands of stars, nebulae, galaxies and star clusters, not to mention the positions of all the known planets. Some of them are even designed to direct motors on the telescope to move it to any target you punch up on their digital displays.

Setting circles and their computerized cousins certainly take a lot of the confusion, trials and tribulations out of finding celestial objects, especially for the beginner. But many purists shun the use of these devices. They prefer to use star maps to hop from one known star pattern to another and home in on their targets by hand. This method, called star-hopping, requires a working knowledge of the major constellations, patience, and a good star chart. It rewards the dedicated observer with greater satisfaction at having found an object and leads to a much more intimate knowledge of the sky. It's amazing what unexpected treasures you may serendipitously stumble across as you star-hop around the sky.

Celestial Road Maps

Just as it's wise to pick up a good road map before you start driving to some city you've never been to, it's even wiser to have a good star map before you set out to explore the heavens. There are no direction signs in the sky and the service stations are few and far between. Fortunately there are a number of excellent star atlases to guide you on your way. If you aren't familiar with the principle constellations, it's worthwhile spending the time to learn them. One of the best ways is to purchase an adjustable star map called a planisphere. It consists of a circular star map that rotates inside an oval window which shows the entire sky for any date and time you set it.

Once you've learned to recognize the major constellations, you'll find using a star atlas much easier. Star atlases vary in price and bulk depending on how detailed they are. The more detailed ones show fainter objects. Turns out the fainter you go, the more objects there are. To show more objects requires a broader scale. This makes for bigger charts and/or more of them as the sky is broken down into smaller and smaller pieces. Choose your first star atlas based on whether or not you can identify the brighter constellations on single charts. If you can't identify the prominent constellations on the atlas, you'll have a hard time orienting them to the real sky. Good atlases for the beginner include the classic *Norton's Star Atlas* (now called *Norton's 2000.0*), the *Mag 6 Star Atlas*, Petersen's *Field Guide to Stars & Planets*, and the *Bright Star Atlas*. More experienced observers will find the *Sky Atlas 2000.0* very useful. The ultimate in a detailed atlas for the amateur astronomer is the *Uranometria 2000.0*.

Another useful reference to have is a guidebook that lists the interesting sights for the celestial tourist to see. Some star atlases like *Norton's* will include such lists as supplements to their star maps. Other atlases will have separate published catalogs of the objects they contain. One of the most useful tour books of the sky for the amateur astronomer is the three volume *Burnham's Celestial Handbook*. It's an ideal companion to a star atlas and one which few, if any, amateur astronomers ever out grow.

MAGNITUDES

Star brightness is measured in magnitudes. The lower the magnitude number, the brighter the star. The brightest stars are magnitude 1 or 0. A few are even bright enough to have negative magnitudes. Each whole magnitude step represents a brightness difference of about 2.5 times. For example, a 1st magnitude star is (2.5 x 2.5 x 2.5 x 2.5) or about 40 times brighter than a 5th magnitude star. On a good dark night the human eye can see stars as faint as 6th magnitude. A good six inch diameter telescope can see stars as faint as 13th or 14th magnitude in a very dark sky.

Seeing the Light - Observer's Flashlights

Inevitably you'll need a flashlight or two to help you with telescope setup, chart reading, and note taking. There are several key things to keep in mind when choosing a flashlight. First, get one just bright enough for the job. A portable search light using 20 D cells and pumping out 10,000 candle power is needless overkill. Furthermore, too much light spoils your night vision.

One important trick that astronomers use is to cover their flashlights with a red filter. It turns out that red light is the least detrimental to your night vision. So get a flashlight just bright enough for your dark adapted eyes to see what you're doing and cover it with red cellophane or red tissue paper. In a pinch, a layer or two of brown paper bag will also work. A good source of red filter material are red tinted, transparent plastic report covers available in many office supply stores. You might also consider having two flashlights, one just bright enough to read your star charts and take notes, and a slightly brighter one for setup and take down of your telescope.

DARK ADAPTATION

When you enter a dark environment your eye starts to adapt to the darkness by opening its pupil wider and by the retina becoming more sensitive to faint light. This dark adaptation is vital to being able to see faint objects in the night sky. Most dark adaptation takes place in the first 20 minutes in the dark, but small gains may continue for hours. Exposure to any bright light source can spoil this dark adaptation. That's why you want to avoid exposure to any bright lights when you observe, including your own flashlights. Some observers use an eye patch over their best observing eye whenever not looking through the telescope. This helps to keep one eye as dark adapted as possible while the other eye is used to examine star charts, make notes, etc. Eye patches can often be found at medical supply stores or costume shops.

CHAPTER 4

Caveat Emptor
Smart Shopping for Telescope Components

Just as a sharp audiophile would shop carefully for the components of a new stereo system, a smart telescope builder is cautious when shopping for components to build a telescope. Sadly, there is no equivalent of *Consumer Reports* magazine in the amateur astronomy world. There are sources of information but they're much more diverse and far less complete.

Two sources of good information about telescopes and components are the dominant magazines in the field, *Sky & Telescope* and *Astronomy*. Both often carry reviews of new products. You can find back issues at any good local library.

An often overlooked resource is your local amateur astronomy group. All large communities and most small ones have these clubs of people from all walks of life who find astronomy a rewarding hobby. Among their members you may well find someone who's built a telescope just like the one you're planning. Others perhaps have bought the particular 'widget' you're considering. These people are usually quite happy to share their experiences with you. Most such groups sponsor regular 'star parties' which are observing sessions where members gather with their telescopes to observe for an evening. Star parties are a golden opportunity to look through other people's telescopes, both commercial and home built. You'll also get some first hand experience on how objects should look through telescopes of various sizes and with different eyepieces. Local astronomy groups are listed once a year in *Sky & Telescope* magazine. Your local planetarium or science museum may also be able to put you in touch with them.

If local star parties are the county fairs of amateur astronomy, the world's fairs are the huge annual gatherings with names like Stellafane, the Riverside Telescope Maker's Conference, the Winter Star Party, Astrofest, and the Texas Star Party. These events draw hundreds of amateurs from all over the country. Here you'll find dozens of commercial and homemade telescopes to examine and even flea markets full of great bargains on used or discontinued telescopes and components. These gatherings are regularly listed and described in both *Sky & Telescope* and *Astronomy* magazines.

If you have a modem equipped computer and access to the Internet or one of the commercial information services like Compuserve, Genie, America Online, or Prodigy; then you have another helpful resource at your fingertips. Most of the commercial services have special areas devoted to all sorts of interests, including astronomy. Many Internet service providers allow access to Usenet newsgroups including 'sci.astro' and 'sci.astro.amateur'. If you take the time to sift through the many messages posted in these areas, you'll find a gold mine of experience about various telescopes and components. If you don't see your item of interest mentioned, post your own message requesting feedback from others. A FAQ (Frequently Asked Questions) file about purchasing amateur telescopes is regularly posted on sci.astro and sci.astro.amateur. It's filled with information about telescopes, components and buyer experiences with mail order dealers.

Once you know what you want, where do you go to get it? Unfortunately few communities have the benefit of a well stocked telescope store. If you're lucky, you may find a selection of commercially made telescopes and a few components in some of the better camera stores or a few of the retail stores that specialize in science related merchandise. Sadly, there's not enough volume and profit in most telescope components to be worthwhile to the average local retailer. Because of this, most telescope components are acquired by mail order.

Most, but not all, sources of telescope components advertise in *Sky & Telescope* and *Astronomy* magazines. You'll also find many of them in the Resource List at the end of this book. Write to vendors for their catalogs or call them if you have questions about their merchandise. If you decide to order by mail, be sure to check on delivery times. Some suppliers may be back ordered as much as a year on popular items. It's also wise to purchase by credit card since this gives you additional leverage when trying to resolve complaints about defective or undelivered merchandise. A reputable company will not bill your credit card until they ship an item to you.

If you're on a tight budget, consider buying used components. You can save some money buying used, if you know what you want and, especially, if you can check it out before you part with your money. On the other hand, there's always some risk in buying used. Usually you can't return a used item if it turns out not to

be what you want or need. Used equipment is advertised in both *Sky & Telescope* and *Astronomy* magazines, but far more ads are found in a newsprint monthly called *The Starry Messenger*.

In the 1950's and 1960's many amateur telescope makers found bargain priced optical components sold as government surplus by companies such as Edmund Scientific and Jaegers. Suitable surplus optics are harder to come by today, but they can be found. If you can find surplus optics in good condition that fit your needs, they can be excellent bargains.

When shopping for optical components look for lenses and eyepieces that are coated. Better yet are optics that are fully multi-coated.

Telescope mirrors are often rated in fractions of a wavelength of light. This refers to the accuracy of the curve of the mirror. In theory, for example, a mirror advertised as 1/8th wave has no deviations from the ideal curve for that mirror greater than 1/8th of a wavelength of light. The smaller this fraction is, the better the mirror and the images it will form (e.g., 1/8 wave is better than 1/4 wave). Regrettably, there is no industry-wide standard on how this measurement should be made, so it's almost impossible to make valid comparisons between manufacturers. Nonetheless, don't consider any optics that are advertised as anything worse that 1/4 wave.

Telescope mirrors can be made of various materials. Plate glass has often been used. It's not very desirable because it expands and contracts with changes in ambient temperature. The precise shape of the mirror needed for sharp viewing temporarily distorts from this expansion and contraction. Mirrors made of Pyrex™ glass expand and contract very little with temperature changes. This makes Pyrex™ the preferred mirror material for amateur telescopes.

Some Assembly Required
Tools, Techniques, & Materials

Safety First and Last

It's far too easy to take safety for granted, especially when working with hand and power tools. Read the following tips and always keep them in mind as you build your telescope.

Make sure you have a good work space. It should provide plenty of elbow room and be free of unnecessary clutter. The floor should be free of any slip hazards and any obstacles that you might trip over. The work area should be well ventilated and have plenty of good lighting.

If you'll be using electrical power tools make sure that the outlets are rated for the amperage your tools will draw and that they are well grounded. Ideally the circuits will be wired with ground fault interruption. Only use grounded extension cords of the proper rating and then only when necessary. Always unplug the tools when changing blades or bits.

Always wear proper eye protection when working with power tools and whenever using a hammer or chisel. Wear loose fitting clothing and don't wear anything like a neck tie that might get caught in a power tool. Wear sturdy shoes with good traction. When sanding, painting, or varnishing wear a dust mask or air filter.

Always thoroughly read, understand, and follow any and all instructions that come with a tool or material. Keep the work area clean and put away tools and materials when you're finished with them. Have a good first aid kit and a fire extinguisher on hand for emergencies, no matter how experienced the workers. Hopefully you won't ever need to use them.

> **H E L P F U L   H I N T S**
>
> • Read and understand the instructions before you start working.
> • Use the right tool for the job. If you don't have the right tool, buy it, borrow it, or rent it.
> • Use clamps or a vise to hold smaller pieces when you drill, tap, cut, file, sand, or machine them.
> • Measure twice (or more!) and cut once.
> • Don't use dissimilar metals together (e.g. a steel screw with a copper washer).

It's a Material Universe

Telescopes are made of a wide variety of materials. Here are some of the more common ones and some tips on handling them.

The best telescope tubes are very rigid, lightweight, and poor thermal conductors. Rigidity is needed to maintain critical optical alignment. If the heat of your body is easily transmitted into the telescope light path it can distort the image. That's why the less thermally conductive the tube, the better. The lighter the tube, the easier it is to transport and counter balance. Heavy walled cardboard has been used for smaller refractors and reflectors. For satisfactory results these tubes must be well sealed to keep moisture from soaking into the cardboard and ruining the integrity of the tube. Phenolic tubes are light weight and poor thermal conductors but they can be quite brittle to cut or drill and need to be sealed before painting. Concrete form tubes (most commonly sold in the U.S. under the brand name Sonotube) are often used. They usually come coated with wax which has to be removed before you can seal and paint the tube. The thin wall of this tubing may need to be reinforced for larger diameter telescopes.

Fiberglass is an excellent tube material, but it must be thick enough so that the tube doesn't flex. Metal tubes, especially aluminum ones, are common and can be very lightweight and stiff. Their thermal conductivity, however, can easily transmit your body heat into the light path of the optics. For this reason, metal tubes are often lined with a layer of thin sheet cork, or other insulation. Small, short refractors can also use PVC tubing so long as the wall is thick enough to prevent flexure. Square or polygonal tubes made of wood are also common.

Telescope tubes should eliminate scattered light as much as possible. This is done by making the interior surfaces as rough as possible and painting them with the flattest black paint you can find. Cylinders of thin sheet cork painted flat black can be cut to size and inserted in larger tubes. Sheets of coarse sand paper, also painted flat black, work well in smaller tubes such as those used on finders.

Telescope mounts can be made from many materials including stainless steel, cast iron, PVC and other plastics, brass, aluminum, and wood.

When using plywood choose an exterior grade, at least grade B if you're going to paint it, or grade A if you plan to stain and varnish it.

Avoid using nails and even screws that can work themselves loose from use. Use bolts and special fasteners like "tee" nuts and right angle connectors to join pieces of plywood. Whenever possible attach wood by clamping it between two pieces of metal using bolts and nuts that allow you to re-tighten the joint should it ever loosen.

Since observing and storage conditions are often damp, be sure to fill all cracks and voids and seal all wood surfaces with a penetrating primer or wood sealer. When painting use the toughest paint you can find. Polyurethane floor paint is a good choice.

Use at least two coats. Be sure to use paints, sealers, varnishes, etc. in a well ventilated area away from any open flames.

When using glues, at the very least use water resistant ones like carpenter's yellow glue. Better yet, find a waterproof glue such as the resorcinol ones. These glues need to be prepared just before use and applied within a few hours of preparation. The additional work is worth it.

Builders of alt-azimuth mounts like the Dobsonian have found that PTFE plastic, such as Dupont's Teflon™, riding against laminate plastic (e.g. Formica™) makes an ideal bearing combination. The goal here is to have a mounting where the observer can easily and smoothly turn the telescope to the desired angle and have it stop precisely where released. Not all plastic laminates work well. Avoid ones with slate-like textures or those with surfaces like embossed tile. A favorite Formica™ laminate among telescope makers goes by the name "Ebony-Star". Scrap Teflon™ and Formica™ sink cut outs can often be found at local dealers. Check your local yellow pages for plastics and laminate dealers.

Often a telescope needs counterweights to achieve proper balance. A frequent source of economical counterweights are old barbell sets. Large fishing sinkers and coffee cans filled with concrete can also work quite well. One enterprising amateur filled the loops of war surplus ammo belts with large bolts to serve as a counterweight belt that could be strapped around his telescope tube wherever needed for the correct balance. Very precise balance can be achieved with a heavy duty, resealable plastic bag filled with the right amount of steel gunshot obtained from a local gun supply store. Another adjustable system employs diving weights or wrist weights from a sporting goods shop attached with Velcro™ to a Velcro™ strip running the length of the telescope tube. Simply attach the weight along the strip wherever the best balance is obtained.

Care & Feeding of Your Cyclops
Telescope Adjustment & Maintenance

With proper care and upkeep, your telescope should last a lifetime. For the most part maintenance is simple and requires a minimum of time, just the sort of thing to do on those occasional cloudy nights when you can't observe.

To keep a Newtonian telescope performing at its peak, you'll need to keep it collimated. This is the most frequent routine maintenance item. Collimation means tweaking all the adjustable optical components so that the light path goes where it's supposed to. Begin by looking down the tube from the open end to make sure that the main mirror and secondary mirror are physically centered in the tube. Next remove the eyepiece and look straight into the holder *(Figure 19)*. You should see the diagonal mirror. Its outline should be centered in the outline of the eyepiece holder and reflect a concentric image of the bottom of the telescope tube. If not, adjust the secondary until eyepiece holder, diagonal mirror, and tube end are concentric. Most commercially made secondary holders have adjustment screws for this purpose. To minimize the chances of dropping a screw driver on the fragile main mirror, make these adjustments with the tube in a horizontal position. The next step is to adjust the screws on the back of the main mirror's cell until the reflection of the secondary and its holder are also concentric as seen through the eyepiece holder. This is best done with an assistant slowly adjusting the

screws one at a time while you observe the effects at the eyepiece holder and direct the adjustments. This method should get you close to correct alignment. Fine tuning should be done with a bright star on a steady night.

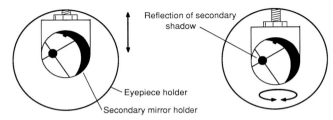

A. First center secondary mirror in holder.

B. Then rotate secondary until main mirror reflection is centered.

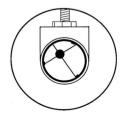

C. Finally, adjust main mirror until reflection of the shadow of the secondary is centered in the diagonal mirror.

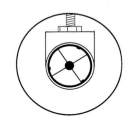

D. Properly collimated.

Figure 19

Refractors seldom, if ever need to be collimated. Most, in fact have no adjustments for collimation. Schmidt-Cassegrains usually just have three adjustment screws on the secondary mirror mount, the alignment of the main mirror being fixed in place. Collimation of a Schmidt-Cassegrain is best made using a bright star on a

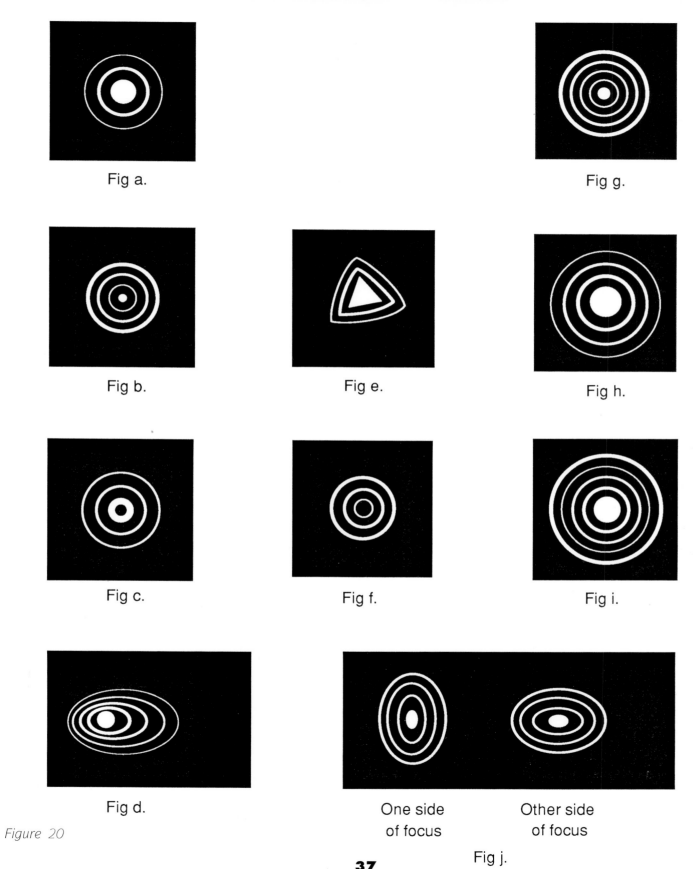

Fig a.

Fig g.

Fig b.

Fig e.

Fig h.

Fig c.

Fig f.

Fig i.

Fig d.

One side
of focus

Other side
of focus

Fig j.

Figure 20

steady night.

To check the collimation of your telescope first choose a star of first or second magnitude located at least 45° above the horizon. Center the star in a low power eyepiece and then increase the power to about 40 to 50x the diameter of the objective in inches (1.6 to 2.0x diameter in millimeters) and focus the star for maximum sharpness. If you now look carefully you should see the star as a round dot of light surrounded by thin faint rings of light. The spot of light is called an Airy disk and the rings are called diffraction rings. Ideally the image should look like *Figure 20a*. Now adjust the focus outwards a little bit until you get a star image with a small dot in the center surrounded by three or four wide circular bands. If you have a Newtonian or a catadioptric you'll see the silhouette of the secondary in front of the star image. If your telescope is properly collimated, you should see an image like *Figure 20b* or *Figure 20c*. If the circles of light are off center or egg shaped, as in *Figure 20d*, the collimation needs adjustment. Making collimation adjustments at high power can be tricky since even a minor adjustment will make a big change in the appearance of the star.

The star test is also a wonderful diagnostic tool for detecting other problems. *Figure 20e*, for example, shows the star image of an objective that's held too tightly in its holder. A mirror with a turned down edge will show a pattern like *Figure 20f*. Over and under correction are shown in *Figures 20g* and *20h*, respectively. *Figure 20i* reveals zonal errors and a pattern like *Figure 20j* betrays astigmatism, if the elliptical pattern rotates 90° when the eyepiece is adjusted from outside to inside of focus. If your optics suffer from astigmatism, zonal errors, a turned edge, or over/under correction, they should be returned to the manufacturer, if possible, for correction or replacement.

Clean Machines

The best route to clean telescope optics is to keep them from getting dirty in the first place. This is done by keeping the objectives and eyepieces covered with snug fitting covers when not in use. Don't throw away any caps that came with your optics!

Incidentally, the plastic canisters, or their caps, used for rolls of 35 mm film make fine 'dummy' eyepieces to plug up 1 1/4 inch eyepiece holders.

Inevitably, some dust will settle on objectives and eyepieces. If there's only a slight amount of dust, leave it alone. A little bit of dust won't degrade the telescope's performance very much, but scratches from hasty cleanings will cause problems. Unlike dust, scratches are permanent. When the amount of dust gets too high, gently blow it away with a can of compressed air designed for lens cleaning. When using compressed air always test spray on a piece of paper first to make sure there's no traces of moisture in the spray. You can usually find cans of compressed air for optics at a good camera store. You can also brush the dust away using very gentle strokes of a clean camel's hair brush. Make sure the ends of the brush hairs are natural rather than cut. Avoid pressing down with the brush any more than necessary to get the dust specks to move and brush them along the shortest path off of the lens or mirror.

If the grime on your lens or mirror can't be blown or brushed away, you may have to resort to cleaning the optical surface with a solvent. Do not use ordinary glass cleaners. They may contain chemicals which could attack the delicate coatings of your optics. Special cleaning solutions designed for camera and telescope lenses can sometimes be obtained at good camera shops, or you can mix up your own solution. Just prepare a 3 to 1 mixture of distilled water and isopropyl alcohol and add a drop of the mildest liquid soap you can find per quart of the solution.

In cleaning a refractor objective or eyepiece, do not remove them from their cells. Eyepieces and multi-element objective lenses are assembled in dust free environments and in precise orientations. If you take them apart you risk getting contaminants between elements and reassembling them in the wrong orientations. Instead, after blowing or brushing away loose dust, moisten clean, sterile cotton balls or cotton swabs with the cleaning solution mentioned earlier. Gently wipe the dirty part of the lens with the moistened cotton ball or cotton swab using only enough pressure to dampen the crud or lens smear. Wipe in a straight line, working from the center of the lens out to the edge. Use a new, clean cotton ball or cotton swab for each wipe. This avoids scratching the lens with a sharp piece of crud picked up from any earlier wiping. If, after the lens dries, any smearing remains give the smeared areas a final wipe with a clean cotton ball or cotton swab soaked only in isopropyl alcohol.

Newtonian objective mirrors are best cleaned in a sink. First carefully remove any rings or loose jewelry that might scratch the mirror. Next place a clean, soft towel in the bottom of a very clean sink. Don't clean the sink with any abrasive cleaners!

Now carefully remove the mirror and cell from the telescope and then the mirror from the cell. Place the mirror, reflective side up, on the towel and fill the sink with room temperature water until the surface of the mirror is under about 1/2 inch (1.27 cm) of water. Add a couple of drops of very mild liquid soap to the sink water. Now take balls of sterile cotton and gently wipe the surface of the mirror under water. Again, wipe in straight lines from the center of the mirror out to the edge. Discard the cotton balls after each wipe and use only the lightest pressure-just the weight of the cotton ball itself is usually enough. After cleaning, rinse the mirror first with cold tap water and then thoroughly with distilled water. Let the mirror air dry in a dust free place.

Even the best coated mirror will eventually need to be re-aluminized. Depending on the environment the mirror is exposed to , this is normally needed every five to ten years for a Newtonian. Re-aluminizing requires special equipment not usually available to the amateur. When the reflective coating of your mirror starts to look dull, have pinholes in it, or you can see through it, it' s time for re-aluminizing. Carefully pack it up along with the secondary mirror and send it off to one of the companies who advertise in *Sky & Telescope* and *Astronomy* magazines.

Other Maintenance

If your telescope has any wood parts it's best to keep an eye out for damage to the paint or varnish. Repaint or re-varnish these nicks and dents as soon as possible to keep moisture from seeping into the wood and warping it.

If you have any ferrous metal parts on your telescope keep an eye out for rust. Eliminate rust as soon as it shows up and consider coating that part with a rust inhibitor.

Use lubricants sparingly. Excess lubricant can migrate to optical surfaces with disastrous results. Check screws and nuts every now and then to make sure they haven't loosened up with time and use.

Orion Nebula, photographed by R. Miller with the 24-inch telescope of the Michigan State University Observatory

CHAPTER 7

Observing 101
An Introduction to Using Your Telescope

Once you've finished building your telescope you'll be itching to turn it loose on the cosmos, an event astronomers call 'first light' for a telescope. Before you zoom off down the Milky Way at light speed, take a few minutes to look over the following tips. They'll help you get the most out of your sky watching adventure.

Check it out in Broad Daylight

It's a good idea to take a shakedown cruise in broad daylight. Start by setting up your telescope in a spot where it will have a clear view of one or more distant earth objects like telephone poles, TV or radio broadcast towers, or high voltage towers. They should be at least a half a mile (0.8 km) away.

Aim the telescope at the distant object by sighting along the side of the tube. Insert your lowest power eyepiece (i.e. the one with the longest focal length) into the eyepiece holder. Slowly move the telescope around by small increments while you look through the eyepiece until you see something other than a gray or blue disk of light. Most likely you'll first see fuzzy blobs of dark and light. When you do, stop moving the telescope and start slowly adjusting the focus of the eyepiece in one direction. If the image seems to be getting sharper, or smaller, keep adjusting the focus in that direction until the image is sharply focused. If the image is getting bigger and/or fuzzier, reverse the direction of your focus adjustment.

Be careful that you don't move the eyepiece too far in or out.

When you've got a focused image you'll notice that it's upside down, or inverted. If you're using a star diagonal it will also be reversed left to right. This is normal for an astronomical telescope. You'll soon get used to it and, after all, in space who's to say which direction is truly right side up? If you plan on doing a lot of terrestrial viewing with your telescope, you might consider an accessory called an *image erector* which will give you a properly oriented image.

Once you've got a distant object in focus, practice looking through the eyepiece. Get used to where your head should be to see the maximum field of view provided by the eyepiece. Learn to look through the telescope without touching it. As you'll discover, even the slightest touch can jiggle the image, or even bump the telescope off target. If necessary, you can sometimes gently rest the side of your nose against the eyepiece to steady your head. Practice slowly adjusting the focus of the eyepiece.

Next repeat the procedure with all of your other eyepieces, working from lower (longer focal length) to higher (shorter focal length) power. Notice how the field of view, eye relief, and image brightness change as you increase power.

After you've had some experience with

adjusting eyepieces it's time to align your finder. Make sure you have your lowest power eyepiece installed and focused in the main telescope when you start this process. Aim the telescope at the distant object and move the telescope until an identifiable point on that object is centered in the field of view of the main telescope. Now look through the finder scope. With any luck, some part of the distant object will be visible in the finder's field of view. Locate the point on the object that the main telescope is centered on in the finder and note where it lies relative to the finder cross hairs. Your goal now is to carefully adjust the little screws that hold the finder in its brackets until the cross hairs line up with the point you've centered in the low power eyepiece of the main telescope. This will take some experimenting to get used to. Just remember not to over tighten the adjustment screws, lest you damage the screws or the finder, or leave them so loose that the finder can fall out of the bracket. Be careful also not to bump the telescope. This will move the center point out of the center of the main telescope's field of view. Keep checking the main telescope to make sure the object is still centered. Once you've got the cross hairs aligned with the low power eyepiece you can fine tune the finder alignment by repeating the process using a higher power eyepiece, but that's really not critical for casual observing. You may find that you've got to realign the finder every time you move the telescope from one location to another. At the very least, you should check the alignment before each observing session.

Once you've familiarized yourself with your telescope during the day, you'll be ready for the main event - night time star gazing! The best night for such an event is any clear night around first quarter Moon.

To the Moon

The first object for your new telescope should be the Moon. It's bright, easy to find, and looks much like the pictures you may have seen in books and magazines. Use your lowest power first. That may allow you to see the whole object at once. You'll notice that there are two major terrain types on the lunar surface: smooth, dark gray areas called *maria* (Latin for seas) and more rugged areas that are white in color, called the *highlands*. Superimposed on both areas you'll see bowl-like depressions of all sizes . These are the famous craters of the Moon. Most lunar craters are named for famous astronomers and scientists of the past. A closer look reveals mountain ranges and long sinuous valleys called rilles. Get yourself a good lunar map and you can spend a lifetime exploring the features of our nearest celestial neighbor. You'll find that surface details stand out best where the Sun is at a very low angle. This border region between darkness and light is called the *terminator*. The jagged line of the terminator slowly moves across the lunar surface day by day as the Moon makes its monthly orbit around the Earth. This gives you something different to look at on the Moon almost every night of the month. At full Moon, the terminator isn't visible since it lies around the circumference of the lunar disk. Unless you like looking at maria, the full Moon phase is one of the least interesting times to observe. When full, there are no long shadows visible on the surface to accentuate the mountains and craters. And the glare from the full Moon makes other observing more difficult.

Before going any further, there are a few important things you should know. First, unless you have setting circles on your telescope and know how to use them, you should take the time to learn to recognize the prominent star patterns or constellations. These will be vital skymarks (the celestial equivalent of landmarks) which will guide you to the objects that are described below.

Second, except for the Moon, don't expect celestial objects to look like the pictures you may have seen of them in books or magazines. Most books and magazines opt for the most exciting and dramatic pictures that they can find. Consequently, they most often use long exposure photographs taken with giant observatory telescopes or from NASA space probes. Often these photographs are enhanced to bring out details. Some even have false color added. Most celestial objects are much fainter, lower in contrast, and lower in detail than those photographs. Nature is much more subtle than the media would like. Often you will not see any color in an object. One reason for this is that the color receptors in the retina of the human eye require more energy (i.e. brighter light) to function. At low light levels like those seen in most small and medium size telescopes, only the black & white receptors work well . The point is, don't have preconceived notions of what astronomical objects are going to look like. In many cases, you may be very lucky to see certain objects at all, much less to see contrasty details or color in them.

With experience, however, the human eye can learn to see finer and finer detail through the telescope. Your first look at Jupiter may only show one or two cloud belts, but with some experience you begin to see more and more fine features. Patience and practice pay off.

Using Averted Vision

What you can see through a telescope and how good it looks depends on many factors,

The human eye has its sharpest vision in the part of the retina that surrounds the center of vision. Because of this you can often see more detail on a planet or a faint deep sky object by looking slightly to one side of the object. This is called using averted vision. Practice this technique. It will come in handy when you take your telescope and your vision to the limit.

not the least of which are the local atmospheric conditions. Astronomers are concerned with two atmospheric qualities: *seeing* and *transparency*. Seeing is the steadiness of the air you're looking through. When currents of hot air are rising in front of the telescope they cause the image to waver and soften just like looking out a window over a hot radiator causes the view to distort. Seeing is often at its worst right after a cold front has moved through or when you are observing over asphalt pavement in summer and warm house roofs in winter. Bad seeing most degrades higher power viewing. At such times you may have to wait several minutes for the seeing to settle down. When it does, steady viewing may only last for a few seconds before the seeing gets bad again.

Transparency refers to how well the atmosphere lets light through. If the air is polluted, thin cirrus clouds veil the sky, or the humidity is high, you won't be able to see faint objects as well. This junk in the air will also scatter any nearby city lights or bright Moon light, adding to the problem. Ironically, some of the steadiest seeing can come with the stale, muggy air that's very poor in transparency. So you might find your best high power planetary viewing on a humid summer night with terrible transparency. Then, after a cold front comes through, high power viewing will be lousy but the better transparency will allow you to find very faint deep sky objects.

It is also most important for the telescope to be in thermal equilibrium with the outside air for it to perform properly. If you take a telescope that's been sitting in a warm house all day out on a cold winter's night, it may take a half hour or longer to cool off and perform well. On humid summer nights a telescope brought out from air conditioned storage may be cool enough for dew to immediately form on all the optics. So, if your telescope is not stored outdoors, allow time for it to adjust to the outside temperature.

Following the Wanderers

Many of the planets can be seen with the naked eye alone. Without a telescope, planets look like stars, except that they don't twinkle like real stars do. The ancients discovered the planets Mercury, Venus, Mars, Jupiter, and Saturn without telescopes because these odd dots of light seemed to slowly wander around the sky against the background of so-called 'fixed' stars. The word planet, in fact, comes from Greek words meaning wandering star. Because of their orbital motions, planets are constantly changing their apparent locations in the sky. Usually you'll find the planets in one of the zodiacal constellations, but not always. Check a current almanac, *The Observer's Handbook*, or issues of *Sky & Telescope* and *Astronomy* magazines for information about the current locations of the planets.

Mercury. This is one of the more difficult planets to see because it never gets very far from the Sun in the sky. It's best found near greatest eastern elongation (the greatest angular distance from the Sun in the sky) in the evening sky right after Sunset, or near greatest western elongation in the morning sky just before Sunrise. Look for it very low in the sky above the horizon nearest the Sun . Through the telescope Mercury will show a very small disk. At times the disk will show phases like a tiny version of the Moon. Seldom, if ever, can you see any detail on the disk of Mercury. Just finding this planet at all is a small victory.

Venus. Venus also stays close to the Sun, but it can wander farther than Mercury. It too is best seen around greatest elongation's when it will easily outshine any star in the sky. Through a telescope Venus dazzles the eye with its brilliant white disk which also exhibits easily seen phases. The dense perpetual clouds of this planet keep us from ever seeing surface features, unless you happen to have giant radar eyes.

Mars. The red planet has always been of great fascination for amateur astronomers and science fiction authors alike. Unfortunately it's such a small world that it rarely appears very big in a small or medium size telescope. The best times to view Mars are when it's near opposition (a time when it's located opposite in the sky from the Sun), which happens about once every two years.

The most obvious feature on Mars is usually one of its white polar ice caps. If you monitor the polar caps over periods of weeks and months, you can watch them grow and shrink with the Martian seasons. Large dark areas on Mars can also be glimpsed even in telescopes as small as 2 inches (11cm) in diameter. A good telescope with an aperture 4 to 6 inches (10cm to 15cm) or better is needed to begin to see the finer features.

Jupiter. The largest planet of our solar system is also one of the best to observe with small to medium size telescopes. Even a 2-inch (11 cm) objective will show the four brightest moons of Jupiter as star-like companions lined up on either side of the cloudy giant. These moons move fast enough that, at times, you can watch them pass in front of or behind Jupiter in a single evening . The same small telescope will also reveal the prominent dark belts of Jupiter and even the famous Great Red Spot. When Comet Shoemaker Levy 9 collided with Jupiter in July of 1994, dark debris clouds from the collision were visible in telescopes as small as 2.4 inches (6 cm) in diameter. With a good objective 4 inches (10 cm) or more in diameter, a practiced eye can see oval clouds, festoons, and other delicate cloud features. As the weather systems of Jupiter evolve, you can view a different Jupiter each year. Recently, the Great Red Spot has been very subdued and even medium size telescopes have a hard time showing it.

Saturn. Like Jupiter, this is a gas giant planet, so all we see is its upper atmosphere. Unfortunately, high level hazes hide most of the activity in Saturn's atmosphere from our view. A few faint belts and zones may be visible. Occasionally, white spots will appear in the Saturnian atmosphere and may last for weeks or months. Saturn's hallmark is its beautiful ring system. Small and medium diameter telescopes reveal three major segments of the rings. The outermost major ring is the "A" ring. Inside of it is the slightly broader "B" ring. Between the two is a narrow, dark gap called Cassini's Division. Under suitable conditions this gap can be seen through objectives as small as 2 or 3 inches(5 to 7.6 cm). Inside the B ring is the darker and vaguer "C" or crepe ring. Depending on where Saturn is in its orbit you may also be able to see the shadow of the rings on the clouds of

Saturn or the shadow of the planet on the rings. About every 14 years, the rings of Saturn appear edge-on as seen from Earth. When this happens, the rings are very difficult to see for several months. For a few days around the exact edge-on date the rings virtually disappear even in the largest telescopes. Saturn's rings were edge-on in 1995.

Uranus and Neptune. These gas giants of the outer solar system are bright enough to be seen in even the smallest of telescopes, if you know where to look. They are so far away that they only show up as tiny bluish or bluish green disks in even large telescopes.

Pluto. This tiny distant world is only a point of light in even the largest earthly telescopes. You need at least an 8 inch (20 cm) telescope to have a fair chance to see this planet since it only gets to 13.7 magnitude, even at its brightest. Since most star atlases don't even show stars this faint, you'll need a special finding chart to locate Pluto. These are usually published each year in the *Observer's Handbook* and the January issues of *Sky & Telescope* and *Astronomy* magazines.

Comets and asteroids. Our solar system is enveloped by a cloud of cosmic gnats called comets. They're little more than big—five to ten miles in diameter (8 to 16 km)—dirty snowballs of frozen gases mixed with dark dust. Most comets live out beyond Pluto's orbit where they're invisible from Earth. From time to time, one of these dirty snowballs will take a roller coaster ride into the inner solar system. As they approach the Sun and warm from it's energy, they release gas and dust which form a cloud called a coma around the dirty snowball. If conditions are right this cloud can be pushed by sunlight and solar wind into a fuzzy tail which can stretch for millions of miles. A dozen or so comets are discovered every year. Most don't get very spectacular but often can be found with amateur size telescopes. They usually appear as small fuzzy stars that slowly move from night to night against the background of stars. Sometimes faint tails can be seen. Once every few years, a spectacular comet will visit our skies. These can get bright enough to be seen without a telescope and have visible tails which can stretch tens of degrees across the sky. Watch the computer bulletin boards or astronomy magazines for announcements about new comets.

Even the brightest asteroids appear as star-like objects in an Earth based telescope. As with comets, night after night they slowly change their positions against the background of stars. Asteroid finding charts are often printed in *Sky & Telescope* and *Astronomy*.

Solar observing. *YOU SHOULD NEVER STARE DIRECTLY AT THE SUN, ESPECIALLY WITH A TELESCOPE!!* Permanent eye damage can result from even brief naked eye glances at the Sun. Solar eye damage, or blindness, can happen without any sensation of pain.

The safest way to observe the Sun is to use a white piece of paper or cardboard held as a screen a foot or so away from the eyepiece and project the Sun's image onto it. This method will allow you to observe Sun spots and solar eclipses in complete safety.

There are safe solar filters available for tele-scopes. Usually they come in one of two forms. One type uses thin sheets of aluminized Mylar™ mounted in a cell that completely covers the telescope objective. This produces a slightly bluish image of the Sun that's safe to view. Just make sure that the filter has no holes in it and completely covers the objective. Only certain types of aluminized Mylar™ are safe as solar filters, so make sure that any Mylar™ you use has been tested and certified safe. Do not use this or any other filter mounted on an eyepiece.

Some imported telescopes used to come with a so-called solar filter. These eyepiece solar filters are not safe. If you come across one don't use it, destroy it.

The other safe filter consists of a flat piece of glass that's been aluminized and transmits only a small percentage of the Sun's light. As with Mylar™ filters, these should only be used over the telescope's objective.

In solar observing, it's better to be safe than sorry. If you can hold a white card up to the eyepiece and see the Sun's image, the filter isn't safe. If you have any suspicion that your method or filter isn't safe, don't use it!

Whichever method you use, make sure you keep the finder telescope covered. This prevents the finder from burning you and from anyone inadvertently looking through it.

Denizens of the Deep Sky
Celestial objects located beyond the solar system are called deep sky objects. This category includes stars, double stars, star clusters, nebulae, and galaxies. Let's discuss each type in turn, starting with the easiest.

Stars are often shunned by some observers as being too mundane. If you check out the histories, distances, spectral types, and other statistics about a star, it makes observing it far more interesting. Try to perceive the subtle colors that different stars have and compare that to their spectral types. Double stars can provide acid tests for the observing conditions and the optics of your telescope. Have a look at Epsilon Lyra, Albireo in Cygnus, and Beta Andromeda.

If you're looking for something a little more dazzling, go after open, or galactic, star clusters. These conglomerations of hundreds of stars can shine like handfuls of diamonds cast on a sky of black velvet. First class examples of open clusters are the Double Cluster in Perseus, the Pleiades in Taurus, and the Beehive in Cancer.

Globular star clusters are more distant star communities containing tens of thousands of stars in a spherical form. Though physically much larger than galactic star clusters, globulars look smaller in a telescope because they're so much farther away. In smaller telescopes they look like fuzzy cotton balls, with perhaps a few of the brighter stars being resolved. As the telescope aperture increases these gems of the sky are resolved into stellar swarms. Prime examples of globular clusters are M13 in Hercules, M15 in Pegasus, and M3 in Canes Venatici.

Of all the classes of deep sky objects, nebulae are the most diverse in visible forms. These are clouds of interstellar gas and dust reflecting starlight or glowing from the energy of hot, nearby stars. Some nebulae, like M42 in Orion and M8 (the Lagoon) in Sagittarius, are stellar birthplaces where matter is contracting to form new stars. Use your lowest power eyepieces to observe these nebulae. Other clouds, like the M1, Crab Nebula in Taurus, are the remains of stars that have violently erupted as supernova explosions. Yet another group are called planetary nebulae because early astronomers thought that some of them looked like the planet Uranus. In reality they have nothing to do with planets. They are the surface layers of old stars which have erupted into space. Unlike most other deep sky objects, planetary nebulae often appear quite small in a telescope. For the smallest ones you may have to use medium to high power to distinguish them from a star.

M57, the Ring Nebula in Lyra and M27, the Dumbbell Nebula in Vulpecula are fine examples of planetary nebulae.

All the deep sky objects we've discussed so far have been part of our Milky Way galaxy. There are, of course, millions of other galaxies in the universe. Quite a few of them can be seen with small and medium size telescopes. Galaxies usually appear as small, unresolved fuzzy splotches of light. Some have oval or elliptical profiles and others have a spiral structure. The Andromeda galaxy (M31), the Whirlpool galaxy (M51) in Canes Venatici, and the Sombrero Galaxy (M104) in Virgo are among the best galaxies to start with.

Notice that many of the deep sky objects listed above have alpha-numeric designations. The "M" refers to a catalog of a hundred plus deep sky objects compiled in the 18th Century by French comet hunter Charles Messier. Unfortunately most new comets look an awful lot like deep sky objects. To avoid mistaking these false comets, Messier compiled a list of them. Ironically, Messier is more remembered today for his list of deep sky objects than any comets he ever discovered. Inadvertently he compiled one of the best lists of deep sky treasures visible in small and medium size telescopes. Many amateurs make a goal of finding all of the Messier objects with their telescopes. Furthermore, there are numerous other fine deep sky objects not found on the Messier list. They're cataloged in two other lists often explored by many amateur astronomers - the Herschel list and the NGC catalog.

At the end of this chapter is a list of some of the easier to find deep sky showpieces that serve as an excellent starting point for the novice observer.

You'll find it very helpful to know the actual

HOW TO STAR HOP

Star hopping is a valuable skill to develop whether or not you have setting circles on your telescope. The basic idea is simple. Use star patterns that you recognize to lead you, stepping stone fashion, to objects you're looking for.

Take for example, the familiar stars of the Big Dipper. If you pick out the two stars in the front of the bowl of the Big Dipper and connect them with an imaginary arrow. That arrow can point in two different directions. If you follow it in the direction away from the bottom of the Dipper's bowl, it will point to the famous North Star, Polaris. How far do you follow the Dipper's arrow? About five times the distance between the two stars in the front of the bowl that you used to aim the arrow. Armed with a basic knowledge of the constellations and a good star atlas you can use this technique to find all sorts of objects.

Another example employs the handle of the Big Dipper. A star atlas will show that the galaxy M51 is located near the end of the Big Dipper's handle. To find it with your telescope first aim the telescope at the star at the end of the handle. Slightly more than 2 degrees west of that star is a fainter star. Find this second star and center your telescope on it. Now aim the telescope about 2 degrees south and a bit farther west. It should now be aimed very close to M51. If you don't see it right away, move the telescope around this spot by very small amounts. In star hopping it helps to use geometric patterns of stars as stepping stones to the objects you're looking for. It also helps to pay attention to the brightness of the guide stars you're using. Usually the brighter a star is, the bigger its dot on the star atlas will be.

field of view of your finder and each eyepiece that you use with your telescope. To get a rough idea of the field of view, at least for the finder and lowest power eyepieces, aim the telescope at the full Moon. The full Moon subtends an angle of just about 1/2°. If you can just fit the full Moon in your lowest power eyepiece, then that eyepiece has a field of view of about 1/2°. If you can fit 10 full Moons across your finder's view, then it has a field of about 5°. To get a more precise measure, locate a star on or near the celestial equator. Such a star will have a declination of close to 0°. The northernmost star of Orion's belt is a good one to use. Aim the telescope so that this star drifts across the widest part of the field of view. Using a stop watch, time how long it takes the star to drift across the field of view. Divide the time (in minutes) by 4 to get the field of view in degrees. For example, if it takes 8 minutes for the star the cross the widest part of the field, the field of view is 8/4, or 2 degrees wide.

Once you have the actual fields of view of your finder and eyepieces, you can use the scale of your star atlas to draw circles representing the fields of your eyepieces and finder on a piece of transparent acetate. Use this overlay as a guide to what the star patterns you see on the atlas will look like through your finder and telescope.

Dealing with Dew

If you observe anywhere that humidity levels get high enough for dew to form at night you'll have to cope with dew covered optics, sooner or later. Refractor objectives and catadioptric corrector plates are the most prone to dew problems, Newtonian mirrors the least susceptible. The best way to handle dew is to prevent it from forming on the optics in the first place.

Aurora Borealis, photo by K. Wilson.

Comet West, photo by K. Wilson.

One way to prevent dew is to keep the telescope optics as dust free as possible. You can also make a dew cap. This is a cylinder, open at both ends, which attaches around the telescope objective lens (refractor) or correcting plate (catadioptric). It should be flat black on the inside and extend a distance of at least 2 1/2 to 3 times the diameter of the objective. Several companies sell resistive wiring devices that are designed to electrically keep the objective or corrector plate above the temperature at which dew forms. Some amateurs also use battery operated hair dryers to remove dew from optics. Bringing a telescope inside a warm car or house can also remove the dew. In any case, don't cover a dewed up lens or mirror until the dew is gone. Dewed optics that have been covered tend to form spots on their surfaces when the dew evaporates. Dew combined with deposits from air pollution can form acids that will etch optical surfaces. Dew can also corrode metal parts.

Backyard Science

Astronomy is one of the few sciences where the amateur, using very modest equipment, can make observations that contribute to the advancement of our knowledge. Observers from all walks of life make magnitude estimates of variable stars; search for new comets, novas, and supernovas (many of which are discovered by amateurs); monitor Sunspots and meteor showers; keep watch on planetary features; and time occultations of stars by the Moon, planets, and asteroids. There just aren't enough professional astronomers or professional telescope time to cover all the available objects and phenomena. You can spend a very satisfying lifetime just exploring all the wonders of the night sky. But, if you ever want to go a bit further and make real contributions to the science of astronomy, the opportunity is there. Just contact any of the following organizations:

- *American Association of Variable Star Observers*
  25 Birch St.
  Cambridge, MA 02138

- *American Lunar Society*
  P.O. Box 209
  East Pittsburgh, PA 15112

- *American Meteor Society*
  Dept. of Physics-Astronomy
  SUNY-Geneseo
  Geneseo, NY 14454

- *Association of Lunar and Planetary Observers*
  Dr. John Westfall, Director
  P.O. Box 16131
  San Francisco, CA 94116

- *International Meteor Organization*
  Dept. of Physics
  University of Western Ontario London,
  ON N6A 3K7 Canada

- *International Occultation Timing Association*
  2760 SW Jewell Ave.
  Topeka, KS 66611

- *Sun SEARCH-Supernova Search & Follow-up Group*
  2700 Valley View Ave.
  Norco, CA 91760

Craters of the Moon through a 6-inch Newtonian reflector, photo by K. Wilson.

First quarter Moon, photographed by R. Miller with a Meade 10-inch f/6.3 Schmidt-Cassegrain telescope.

Waxing gibbous Moon, photographed by R. Miller with a Celestron 8-inch Schmidt-Cassegrain telescope.

Sun spots photographed with a special solar filter, by K. Wilson.

Annular eclipse of the Sun, photographed on May 10, 1994 by K. Wilson.

"Diamond Ring" effect during total Solar eclipse, photographed on Feb. 26, 1979 by K. Wilson with an 800mm lens.

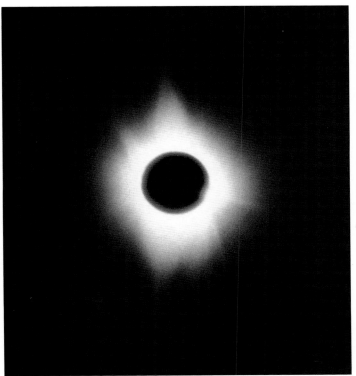

Total Solar eclipse, photographed on July 11, 1991 by Joann Kalemkiewicz with a 500mm lens.

Total Solar eclipse, photographed on July 11, 1991 by R. Miller.

Total Solar eclipse showing middle corona, photographed on Feb. 26, 1979 by K. Wilson with an 800mm lens.

5-hour exposure of circumpolar star trails, photographed by K. Wilson.

1 1/2-hour exposure of circumpolar star trails, photographed by K. Wilson.

M31—the Andromeda Galaxy as it might appear in a finder or small telescope, photo by K. Wilson.

M45—the Pleiades Star Cluster as it might appear in a finder or small telescope, photo by K. Wilson.

A SELECTION OF DEEP SKY GEMS FOR THE BEGINNER

Object	Constellation	Coordinates Right Ascension Hours	Minutes	Declination Degrees
DOUBLE STARS				
Epsilon	Lyra	18	44	+40
Albireo	Cygnus	19	31	+28
Alcor/Mizar	Ursa Major	13	24	+55
Castor	Gemini	7	34	+32
Beta	Andromeda	1	9	+36
Alpha	Capricornus	20	18	-13
OPEN CLUSTERS				
Pleiades (M45)	Taurus	3	47	+24
Beehive (M44)	Cancer	8	40	+20
Double Cluster (NGC 869/884)	Perseus	2	20	+57
M7	Scorpius	17	54	-35
GLOBULAR CLUSTERS				
M13	Hercules	16	42	+36
M3	Canes Venatici	13	42	+28
M5	Serpens	15	19	+ 2
M15	Pegasus	21	30	+12
NGC 5139	Centaurus	13	27	-47
M4	Scorpius	16	24	-27
DIFFUSE NEBULAE				
Orion Nebula (M42)	Orion	5	35	-5
Lagoon Nebula (M8)	Sagittarius	18	4	-24
Trifid (M20)	Sagittarius	18	3	-23
Omega Nebula (M17)	Sagittarius	18	21	-16
PLANETARY NEBULAE				
Ring Nebula (M57)	Lyra	18	54	+33
Dumbell Nebula (M27)	Vulpecula	20	00	+23
Saturn Nebula (NGC 7009)	Aquarius	21	4	-11
Ghost of Jupiter (NGC 3242)	Hydra	10	25	-19
GALAXIES				
Andromeda Galaxy (M31)	Andromeda	00	43	+41
Whirlpool Galaxy (M51)	Canes Venatici	13	30	+47
M81	Ursa Major	9	56	+69
Sombrero Galaxy (M104)	Virgo	12	40	-12
NGC 253	Sculptor	00	48	-25

Star Party of amateur astronomers, at dusk with crescent Moon and Venus, photographed by K. Wilson.

2.5-inch f/4 rich field telescope built by Jocelyne Trottier of St. Eustache, Quebec, photographed at 1994 Stellafane Convention.

30-inch ultra-light weight telescope built by Steve Overholt, photographed at 1994 Astrofest.

10-inch light weight, tubeless Dobsonian (a variation of Alice *in this book) built by Ron Ravneberg.*

10-inch Dobsonian (like Project 6 in this book) built by Jerry and Susie Stimplfe. Note the extra set of altitude bearing rings added to the mirror support box for fuller rotation of the tube assembly.

A 6-inch solar telescope built by Joann Kalemkiewicz. This novel telescope features a full aperture solar filter, tilted 45 degrees to the optical axis. This filter acts as a beam splitter to reflect most of the sunlight away. The remaining light enters the telescope and is focused by the primary mirror back to the tilted front mirror which serves as the Newtonian diagonal.

A 10-inch, f/5.6 wooden-tube reflector built by Ed Meissler. The tube was formed using 52 wooden slats, 3/4 inch wide, 1/4 inch thick glued along the edges.

Henry Stockmar of Richmond, Virginia with his 8-inch f/6 Newtonian reflector and its unusual plastic sphere alt-azimuth mount, which glides on three nylon furniture glides.

A 22-inch f/5 reflector which has a computer readout of its position. This telescope is outstanding for its workmanship and was built using 3/4 inch Baltic Birch plywood by David Deremo.

Close-up of David Deremo's telescope mirror box and altitude bearings. The two wooden wheelbarrow-like handles allows this telescope to be moved easily by one person.

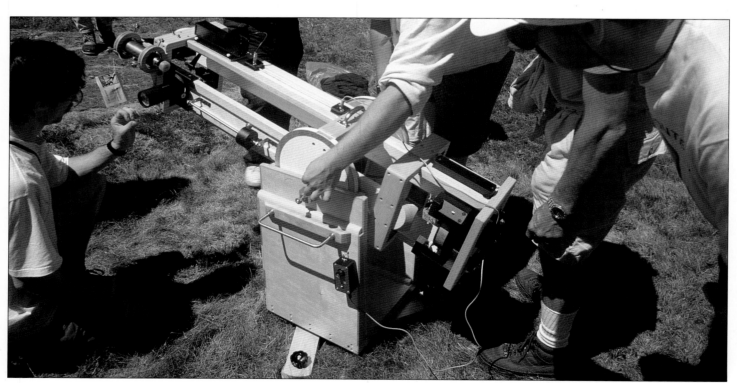

A 5-inch Dobsonian reflector exhibited at the 1994 Stellafane Convention by Jerry Wolczanski of Warrenton, Virginia. Note the wooden framework and open tube.

Photo by David Deremo showing his 22-inch f/5 telescope with an observer's work table. This table has a red back-lighted top surface for reading star maps and has drawers for eyepieces and accessories.

David Deremo's telescope showing the mirror cell and handles for moving.

Comet hunter, Don Macholtz, and his home built Newtonian reflector. Note the pipe equatorial mount and barbell counterweights.

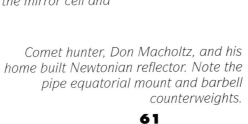

A 17-inch Dobsonian by Tim Baker. This photo shows the mirror cell. Note the wooden blocks outside the mirror box into which the aluminum truss tubes fit.

Tim Baker's 17-inch telescope mirror box with the mirror exposed.

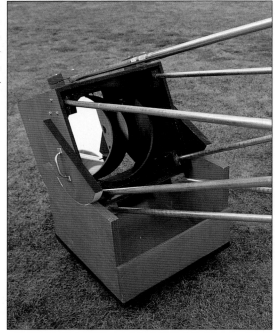

A simple variation on the Dobsonian reflector design. Note the low-to-the-ground rocker box, designed for compactness and stability, and the use of plywood in the rocker box.

A Newtonian reflector with a simple equatorial mount made of plywood and plumbing pipe.

Amateur astronomer with an 8-inch Newtonian reflector on an equatorial mount, photographed by K. Wilson.

Catching Photons
An Introduction to Astrophotography

Sooner or later most amateur astronomers develop a desire to capture what they see on film. Astrophotography, as it's called, can be one of the most rewarding pursuits and one of the most frustrating. At best, one in ten astrophotos taken is satisfactory, one in a hundred is outstanding. This implies a lot of trial and error, and the patience that goes along with it. Although we can't cover every aspect of astrophotography in this book, the following introduction will get you started on the right foot.

Cosmic Cameras

Obviously the prime piece of equipment you'll need is a camera. Although astrophotos have been taken with everything from Brownie box cameras to large portrait cameras, the most popular format, by far, is 35mm. A good 35 mm camera for astrophotography needs the following features: single lens reflex (SLR) focusing; interchangeable lenses; a mechanical shutter with speeds that can be set manually; and provision for the use of a cable release.

Unfortunately many of the 'idiot-proof' 35 mm SLR cameras on the market today have automatic electronic shutters and non-changeable lenses. These cameras were not designed to have their shutters left open for the minutes or even hours needed to capture faint astronomical objects. If you try such long exposures with them you may find that the battery dies long before your exposure is over. If you have one of these 'new fangled, automatic everything' cameras you may be able to take short exposures, say a couple of seconds or less, of bright objects like the Moon or planets. But, if you want to go after anything fainter, you've got two choices . You can track down a used model of one of the older, low tech cameras with a mechanical shutter like the Nikon F, Nikon F2, Canon F-1, Canon FT, Canon FTb, or Olympus OM-1. Alternatively you could get yourself one of the few modern cameras that still have a mechanical shutter, like the Pentax K1000, the new Canon F-1, Contax S2, Nikon F3, Nikon FM2, Pentax LX, or Olympus OM-4T.

Other desirable features in an astrocamera are interchangeable focusing screens and mirror lockups.

Good Gadgets

In addition to the proper camera, you'll need a few other items to get started. A good cable release, for example, will allow you to trigger the shutter without jiggling the camera. A sturdy tripod allows you to take wide and normal angle long exposures of stars, auroras, meteors, bright comets, etc., and some through the telescope exposures. If you are going to do much picture taking through your telescope, you'll need a T-ring adapter for your camera and an 1 1/4 inch adapter to connect the camera to the telescope eyepiece holder. Look for an adapter that also allows you to insert an eyepiece into it.

If you get badly bitten by the astrophotography bug, you'll want to look into other accessories like wide angle and telephoto lenses, special focusing screens, magnifying view finders, off axis guiders, clock drive correctors, illuminated cross hairs, guide telescopes, and so on.

Happy Star Trails to You!

Let's start with the simplest form of astrophotography, star trails. Load your camera with a relatively fast film, ISO 200- 400. Attach a normal or wide angle lens, set the iris wide open, and the shutter speed to "B" or "T" for a time exposure. Attach a cable release and put the camera on a tripod. Aim the camera at the North Star, set the focus for infinity, open the shutter with the cable release and lock it open. If you live in the southern hemisphere, aim the camera south and tilted up at

A 35 mm SLR camera mounted on a sturdy tripod for astrophotography (e.g. star trails, auroras, constellations, bright comets, etc.)

angle equal to your latitude. Leave the shutter open for 30 minutes to 3 hours, depending on how dark the sky is where you are. Carefully close the shutter. When the film is processed, it should show arcs of light produced as the Earth turned on its axis during the exposure. In the northern hemisphere, the bright star with the shortest arc will be Polaris, the North Star. It's the closest bright star to the north celestial pole and exhibits the least amount of motion as the Earth rotates.

If you repeat this procedure, but aim the camera at stars near the celestial equator, like those of Orion, the star trails will be straight, parallel lines. The longer the exposure, the longer the star trails will be. If your exposure is too long, your film too fast, your lens too fast, or the city lights in your neighborhood too bright, the exposure may get washed out by light pollution. For best results experiment with the variables. Remember what we said about trial and error?

An interesting variation of star trail photography is to interrupt the exposure at the end by blocking the lens with a black piece of cardboard for about five minutes, then uncover the lens for 30 additional seconds of exposure and then close the shutter. This will leave a gap in the star trails followed by a round dot of light from each star so that you can more easily recognize the constellations.

As you take star trail pictures, you may find unexpected streaks crossing your star trails at odd angles. These can be caused by artificial satellites, airplane lights, and even meteors.

Now that you've gotten your feet wet, let's take the camera and tripod combination a little further. Instead of letting the Earth's rotation streak out star images, let's shorten the exposures so that we freeze the motion. The proper exposure to stop trailing from showing depends on the focal length of the lens you are using and the part of the sky you are photographing. Stars near the celestial poles move the least over time and the ones near the celestial equator move the most. The longer the focal length of the camera lens, the sooner trailing shows up. As an example, consider photographing the constellation Orion with a 50 mm lens. Since Orion straddles the celestial equator, it will show the greatest motion over time due to the Earth's rotation. At most you can expose your film for about 30 seconds before trailing will start to show on the film. If you were instead aiming at the Little Dipper, which is near the north celestial pole, you could expose about six times longer without trailing showing up. If we used a 24 mm lens, instead of the 50mm, we could double the exposure time without the trails showing.

By using a fast 50 mm or wider lens, a tripod, fast film, and the right exposure, you can build your own photographic atlas of the constellations. The same techniques will allow you to photograph auroras, meteors, lunar halos, the Milky Way, and even bright comets.

The Ultimate Telephoto Lens

Your telescope can be the ultimate telephoto lens. Most amateur telescopes have focal lengths equal or greater than the longest lenses carried by a photographer. As with any telephoto lens, the sharpness of your pictures will depend on the quality of the optics and how firmly you support the lens. The more stable the telescope mounting, the easier it will be to get good photos.

You can mount a camera on a tripod and carefully adjust it until it sees a good image through the telescope's eyepiece, but this is very tedious and succeeds only with quite short exposures.

Rather than putting the camera on a tripod to shoot through the telescope, it's best to get a camera-to-telescope adapter. It will allow you to physically attach the camera to the telescope tube or eyepiece holder. The simplest of these adapters consists of a 1 1/4 inch metal tube threaded on one end to fit a T-adapter. The T-adapter attaches to your camera just as any lens would.

You Gotta Have a System

There are three basic systems of astrophotography through a telescope: afocal, prime focus, and projection. The afocal method entails positioning the camera with its normal lens in place so that it looks through the focused eyepiece of the telescope. The camera can either be mounted on a tripod or a special adjustable bracket attached to the telescope. In either case you have to adjust the camera's position until it's aimed straight into the eyepiece and as close to it as you can get without touching the lenses and still get a focused image in the camera. Both the telescope and camera must be focused for infinity.

Prime focus photography uses your telescope's objective as a telephoto lens without any intervening eyepiece or camera lens. Simply remove the camera lens and replace it with the 1 1/4 inch adapter with its T-adapter in place. This arrangement provides the brightest image and shortest focal length of any telescope camera combination. Prime focus is the simplest technique for the novice astrophotographer to start with. It provides the widest fields, brightest images, and shortest exposure times. It's ideal for large objects like the Moon.

To use the prime focus method make sure that the focuser of your telescope will allow you to get the focal plane of the camera up to the focus of the telescope objective. In some Newtonians you may have to drill a second set of mounting holes in the tube for the mirror cell. Position this set of holes 2 inches(5 cm) closer to the secondary mirror. Then carefully remove the screws holding the cell to the tube and gently move the cell up to the new position. This should allow your camera to focus in the prime focus arrangement.

A variation on prime focus photography employs a Barlow lens between the objective and the lensless camera. This is called negative projection. Negative projection can be very useful where you need a little more magnification than plain prime focus can give you, or to get the focal plane to the camera's film plane without moving the mirror up.

If your camera to telescope adapter has an additional tube for eyepiece projection, you can increase the magnification to a much greater extent. Eyepiece projection can produce the highest magnifications of any astrophotographic system. It's ideal for capturing small details of the Moon and planets. It's also the most difficult system to get an object centered and focused with. Since it produces the longest focal lengths and greatest f-ratios it also requires the longest exposures.

Suffering from Exposure

One of the most perplexing challenges for the novice astrophotographer is determining the proper exposure. Astronomical subjects come in the widest range of brightnesses and no camera meter is up to the job of determining the best exposure.

To determine the right exposure you must first calculate the system focal ratio of your telescope-camera combination. The following formulas will help you to work out this vital number:

For prime focus astrophotography:
$$S = T$$
For negative projection astrophotography:
$$S = T \times BM$$
For eyepiece projection astrophotography:
$$S = T \times \frac{D - E}{E}$$

For afocal astrophotography:
$$S = 2 \times \frac{P}{O}$$

Where: S is the effective focal ratio of the telescope-camera combination; T is the focal ratio of the telescope objective; BM is the magnification factor of the Barlow lens; D is the distance from the center of the eyepiece to the back of the camera; E is the focal length of the eyepiece; P is the magnification of the telescope-eyepiece combination; and O is the diameter of the objective in inches. Note E and D need to be in the same units (e.g. inches with inches).

Once you have the effective focal ratio of your system, plug that number into the following formula to get an exposure starting point:

$$\text{Exposure time (in seconds)} = \frac{S^2}{\text{Film Speed} \times B}$$

Here film speed is the ISO number of the film and "B" represents a brightness factor that varies from object to object. Select the appropriate value for B from the following table:

OBJECT	B VALUE
Quarter Moon	15
Full Moon	64
Venus (brightest)	1024
Venus (at elongation)	612
Mars	15
Jupiter	128
Jupiter's main moons	0.128
Saturn	8
Uranus	0.032

The above formula should only provide a starting point. Variations in atmospheric transparency, altitude of the object above the horizon, and variations in object brightness may cause the actual B values to vary from the table. To allow for this astrophotographers bracket their exposures. This means taking at least one exposure above and one below the recommended value. For example, the quarter Moon photographed with an f/8 system on 100 speed film has a suggested exposure of about 1/15th of a second. To bracket, take additional exposures at 1/8th and 1/30th of a second.

Keep accurate records of each astrophoto you take. Note the date, time, location, object, telescope aperture, telescope f-ratio, eyepiece used (if any), projection distance (if any), seeing conditions, film type and speed. That's the only way you'll progress very far on the astrophotography learning curve.

Generally speaking, astrophotographers prefer to use the fastest film they can get away with without the graininess of the film getting objectionable. As with any photography, the faster the film, the larger and more noticeable the grain of the film emulsion. In the black & white category T-Max™ 400 and Technical Pan™ 2415 are very popular. If you prefer color slides, try Kodachrome™ 200, Fujichrome™ 400, Scotch Chrome™ 400, or Agfachrome™ 1000. For color print film, good results have come from any of the Ektar™ films, Fuji HG™ 400, and Konica SR-G™ 3200.

If you don't process your own film, take it to a good reliable lab. Always take at least one normal daytime scenic shot on each roll of astrophotos you take. This gives the processor a reference frame to set their negative cutter. Otherwise they may cut your prized constellation shot right in half! Many astrophotographers shoot a sheet of paper with their name and address on it as a daylight reference frame. This gives you added security against lost film. Some processors will leave your film uncut if you ask them. That way you can cut the film right where you want to. If you're using print film, it's worth telling the processor that the roll includes astrophotos. It may help them get the exposure of the prints right if they know it's supposed to be the stars of Orion and not the candles on Uncle Joe's birthday cake! Most processors will re-do your prints if you're not happy with them. So don' t be shy about sending them back with a note that the craters on the Moon are not supposed to be green!

Guide to Long Exposures

If you plan on taking pictures through your telescope of anything fainter than the Moon, Jupiter, Saturn, Mars, Venus, or Mercury, you'll wind up

A Celestron 8-inch Schmidt-Cassegrain telescope set up for solar photography. A 35mm camera is attached to the focus at the rear with a T-mount and adapter. A full aperture solar filter covers the front end of the telescope.

taking exposures lasting tens of minutes, or even hours to get an acceptable image. This is due not only to the faintness of the objects but also to an effect of most films called *reciprocity failure*. Simply put, reciprocity failure causes the film to become less sensitive to light over exposures lasting more than a minute or so. To over come this requires longer and longer exposures, and each one needs to track precisely on the object.

This requires a polar aligned equatorial telescope with an accurate clock drive, an electronic corrector for that clock drive, and an optical system to guide the telescope during the exposure. In addition, experienced astrophotographers often treat their films with special gases and chill them with dry ice to reduce the effects of reciprocity failure. All these techniques are beyond the scope of this book. When and if you reach the point that you want to explore them, refer to one of the books in the resource list at the end of this book.

High and Low Tech Imagining Alternatives

Photography is not the only way to make permanent records of the celestial orbs. There are alternatives, from inexpensive and simple to complex and expensive.

At the simple end lies the artistic path. With a simple sketch pad and pencil you can make drawings of the Moon, planets, comets, deep sky objects - anything that your eye can see. The human eye, in fact, can often detect very fine planetary detail that's too fleeting for film to catch. You can also sketch a wider dynamic range of brightness visible in deep sky objects than any film is capable of rendering in a single exposure. Some amateurs use their pencil or charcoal sketches as a basis for full color paintings.

Other hobbyists use video to capture astronomical events, especially things like eclipses. A video camera needs to be very sensitive to low light to pick up even the brightest astronomical objects. Most cameras with an automatic iris overexpose on bright objects like the Moon unless you zoom in enough that the object fills most of the frame. If you have this problem, try setting the shutter to a higher speed or add a telephoto attachment to the lens.

The latest gadget boom in astronomical imaging is the CCD (Charge Coupled Device). These devices are microchips with arrays of tiny light sensitive elements that output a digital image to a computer. Once stored in computer memory, CCD images can be enhanced in ways undreamed of by film photog-

raphers. CCDs have been used by professional astronomers for many years. In recent years they've also taken over the consumer camcorder market. Camcorders, unfortunately are limit to exposures of less than 1/30 of a second. The same sort of chips can be re-configured to store up faint light over exposures lasting many minutes or hours. And, the volume of chips made for camcorders brought price drops that have put them within the reach of the amateur astronomer. CCD still cameras start at about $300. Throw in a computer to store and process the images, a cooling system to reduce electronic noise (from the CCD) and a CCD system can wind up costing as much or more than an astrophotography set up. Although most CCD images require shorter exposures than film, especially for deep sky objects, calibration images and equipment setup make CCD imaging on a par with astrophotography in the hassle department.

An aluminized Mylar™ solar filter mounted on a small Schmidt-Cassegrain telescope.

Project Overview

The projects included in this book cover a range of instruments you can build. By following these directions, you can construct an instrument which will be of lasting value to you, whether you are a newcomer to astronomy, or an experienced observer. These particular projects were selected because they cover several useful designs for reflecting telescopes. Apertures range of 4 1/4 inch to 10 inches, from small telescopes which can be hand-held or attached to a camera tripod, two different equatorially mounted telescopes and a typical large-aperture Dobsonian. All these telescopes are relatively simple, therefore they are affordable and can be built with a small investment in time. We have not burdened these basic projects with elaborate features.

All projects can be completed successfully, even by those relatively inexperienced in woodworking, using common workshop tools. The authors have built many telescopes with only basic tools. It is assumed that you have these tools available: a hand saw (ideally a table saw, even if borrowed), a drill with an assortment of drill bits, hammer, screw drivers and at least one adjustable wrench. A small shop with a workbench or table in your basement or garage is perfect, where you can get away with making sawdust and a little noise. A reasonably clean, ventilated area for painting should be available. (For painting, a workshop where dust can be swept will do, if you let any airborne dust settle overnight.)

Power tools are convenient time savers, but they are not generally necessary for these projects. If you need a tool and can't afford one, consider borrowing or renting it instead. When using power tools, be sure you know how to use them correctly. Safety is always first, so read and follow manufacturer's instructions.

Perhaps you may prefer to work with a friend. Finding some local amateur astronomers and other telescope makers in your area or joining an astronomy club can provide a source of inspiration and assistance.

One topic, dear to a few telescope makers, making your own mirror, is avoided. One of the authors (RDM) has finished about a hundred mirrors for telescopes and other instruments. Mirror- and to a smaller extent, lens-making is still practiced by some builders, but as evidenced by the very few sources for supplies, a decreasing percentage of builders make their own optics. Most others would prefer to purchase the optical systems from one of the many companies who sell good, affordable mirrors. Just a few decades ago, such commercial optics were not so widely available and telescope makers had little choice but grind their own. Patient amateurs who can afford the time and effort can produce some of the very best optical components. Telescopes having such optics will indeed give breathtaking views of the heavens.

If you are interested in astrophotography, the first project will provide you with a star tracker which can be used to take wide-area, time-exposure photographs of the constellations, without the motion blur caused by Earth's rotation. It is easy to use, and fun to build. With exposures of a minute or less with fast (400 speed) film in a 35mm camera with an f/2 or f/2.8 lens will show more stars than can be seen with the naked eye. Longer exposures made during times of meteor showers can produce some striking pictures. Photographing the same part of the sky a few days or a week apart throughout a month or two, will show the fascinating and complex motions of the planets.

Projects 2 and 3 feature two different ideas for a small rich-field telescope. The PVC tube telescope in Project 2 is simple to build. Regardless of any larger telescope you may own, this one will still be handy, because it can be taken outside and used at a moment's notice. Using one of these telescopes on a rigid camera tripod is a great way to view star clusters and galaxies. Project 3 takes the same optical components and mounts them in a beautiful wooden tube telescope, similar to some made a century ago.

Project 4 is a larger rich field telescope, one of six inch aperture on a unique all wooden equatorial mounting. If you prefer a modest size telescope with a portable equatorial mounting, this one is easy to make and is a delight to use. This mounting could be scaled down and used to mount the telescope featured in Project 2.

An equatorially mounted eight inch telescope is featured in Project 5. With the hour angle and declination axes made from hardware store pipe fittings attached to a simple wood support structure, it is a good choice for an economical telescope. The elegant design of the wooden members of this mounting permits this telescope to be setup in a minute with no fumbling with nuts and bolts (and nothing to lose in the grass at night).

A large Dobsonian-style telescope with its simple alt-azimuth mounting is the topic of Project 6. For a large telescope, this one can be built very easily, with readily available parts, in a week or two of evenings. Details of the telescope and mounting are presented here, but the basic design can be scaled up or down to accommodate telescopes of different apertures or focal ratios.

The bonus project, contributed by Ron Ravneberg, is presented as an inspiration. Two of Ron's telescopes were displayed at the Astrofest conference in September, 1994. His telescopes were impressive for their quality of construction, thoughtful attention to design details and innovative ideas. Ron graciously agreed to write an article for this book about his unique, portable eight inch reflector. In this article, detailed instructions have been omitted; the construction techniques are similar to those presented in the Projects. You're encouraged to borrow these ideas and expand on them for your own custom telescope.

With a little patience, and a carefully planned approach, almost anyone can build a telescope which will be rewarding use and a delight to show to others.

The Projects use a few words of jargon, so you can become familiar with the important terms first, here they are:

Primary: the main light gathering optical surface; it can be either a lens or a mirror. For refractors and simple reflectors, it is the first optical surface that the light ray encounters on its way through the optical system. The primary is also called an objective.

Secondary: In a Newtonian telescope, light reflected from the concave primary converges toward a point located near the front of the tube where all rays meet. In order for the observer to view a star, the light must be diverted to a point outside the tube, where an eyepiece is located. This requires the use of another reflecting surface; usually a flat mirror in the shape of an ellipse is used. The shortest line that can be drawn through the center of the ellipse that meets both edges is called the minor axis; similarly, the major axis is the longest such line. In optical catalogs, these mirrors are specified by their minor axis dimension.

Optical axis: an imaginary line drawn through the center of a lens or mirror, perpendicular to the surface, is the optical axis. It is a line of symmetry for the primary.

Aperture: the diameter of a circular lens or mirror which collects and focuses light.

Focal length: A lens or mirror takes parallel rays of light from a star and concentrates all that light at one point; this point is called the focus. The distance between the center of the lens or mirror and the focus, measured along the optical axis, is the focal length.

Focal ratio: the number determined by dividing the primary's focal length by its aperture. For example, a 24" focal length and a 6" aperture combine to produce a focal ratio of (24/6) = 4. In this example, the focal ratio is be expressed by writing it as f/4. Photographic lenses are described in the same way, and focal ratios are often called f-numbers.

Eyepiece: the lens or combination of lenses through which the star's image is magnified.

Exit pupil: the location in space where the light rays emerging from an eyepiece pass through a circle of smallest diameter. The exit pupil size is the diameter of this circle. When the position of your eye coincides with the exit pupil, you will see a fully illuminated field of view.

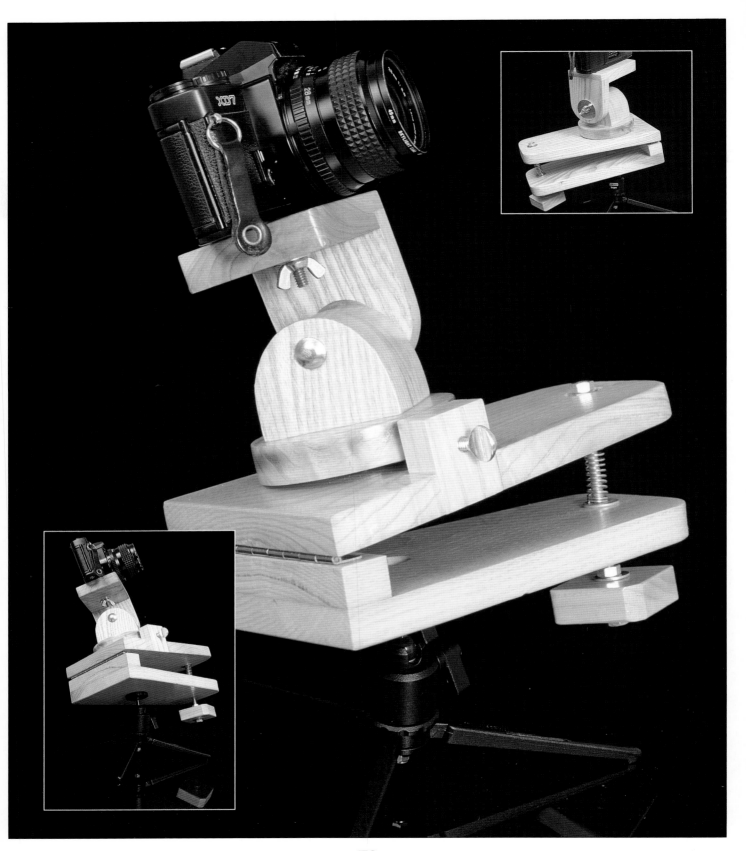

A Star Tracker For Your Camera

The desire to capture night sky wonders with photographs has provided the motivation for our first construction project. It may come as a surprise to see that some interesting and beautiful wide-field constellation pictures can be obtained with simple equipment.

The requirements for basic astrophotography are: almost any camera, such as a 35mm camera, not fully automatic, which has a shutter speed setting for time (T) or bulb (B); a very sturdy tripod and a cable release. Any black and white or color film with a film speed ISO 400 is good for a beginning.

With such modest equipment, you can photograph the sky. On a dark moonless night, try aiming your camera at a favorite constellation and setting the lens aperture to wide open, ususally f/1.4 to f/2 and using the cable release, make a time exposure of 12 to 15 seconds. This exposure time will record most of the stars visible to the naked eye. You may want to try longer exposures if you are fortunate to have a dark sky, relatively free of glaring street lights. To make "star trail" pictures where the stars are trailed because of the Earth's rotation, try aiming your camera toward Polaris. For longer exposures of several minutes to an hour or more, set the aperture from f/4 to f/11 and experiment with different exposure times. When aiming toward Polaris, even a picture with an exposure time of a few minutes will show the stars trails as concentric arcs centered near Polaris. If you are in the southern hemisphere, you will not find a convenient "pole star." The sky appears to rotate around the south celestial pole in the constellation of Octans.

Next, try a few star trail pictures with your camera aimed east or west. The stars will appear trailed at an angle to the horizon. Some interesting effects may be obtained with exposures of several seconds to several minutes of the sky and also include some landscape if a little moonlight is available.

Eventually you may want to take some pictures of the Milky Way or of constellations in which the stars appear as natural points of light rather than trails. To do this, the camera must be moved continuously or (at least once every ten or fifteen seconds) to compensate for the apparent motion of the sky caused by the Earth's rotation.

A very simple and effective star tracker which will do the job can be built for less than ten dollars. The idea is simplicity itself. Since the sky appears to rotate around a point (the celestial pole), a camera needs to rotate to compensate. We will build what is known as a "barn door" tracker which basically consists of two pieces of wood attached at one end with a hinge in the form of a "V". The open end of the "V" contains a threaded rod, which will open and close the two pieces. Imagine the "V" on its side. Attach the bottom side to a tripod and mount a camera on the top piece. Aim the hinge toward Polaris and this will make the hinge parallel to the Earth's axis of rotation. When the threaded rod is turned at just the right rate, the camera will rotate about the hinge and compensate for the apparent motion of the sky. Voilà, pictures of the sky with natural looking stars!

The bottom board surface is fitted with a tee nut. The 1/4-inch x 20 screw on a standard camera tripod goes into the Tee nut, fastening the tracker to the tripod. Your camera attaches to a swivel and tilt head mounted on the top board. Photo 1 shows our own version of a swivel mount. Photo 2 is a tracker built on the same principal by Dick Weiser from an article by Frank Zullo (in *Discovering Arizona's Night Sky*); Dick's mount employs a universal ball joint to attach the camera. Plans for the tracker are displayed in figures 1, 2 and 3. Note the two sides of the "V" are separated at the hinge by a 3/4-inch thick wooden spacer bar. This allows the "V" to close without causing undue spring compression at the opposite end.

The tracker is simple to build and with only one exception, dimensions are not critical. That exception is the distance between the center of the hinge and the threaded rod. With a correct choice of thread pitch and distance from the hinge, a "clock" arrangement exists which causes the tracker to follow the sky as the rod is turned at a steady rate. The rod should be turned at one revolution per minute, inasmuch as this is simply done by looking at the second hand of a watch. (A quarter turn of the rod every 15 seconds is sufficient.) The distance between the hinge and the rod in inches is represented by the formula:

$$d = 228.557/tpi,$$

where tpi is the number of threads per inch. Our tracker uses a fine thread, 28 threads per inch, making it compact — less than 9 1/2 inches total length, where d is 8.163 inches. Only well-stocked hardware stores carry 28 thread per inch, 1/4-inch diameter rod. The more common size is 1/4-inch with 20 tpi, so d becomes 11.43 inches. It is advantageous to shop around for a rod with finer threads, but if you must use a coarser threaded rod, be certain to adjust the lengths and spacing on the plans accordingly.

CAMERA TRACKER PARTS LIST

4) Pieces of sturdy 3/4 thick wood at least 6 by 10-inches. Pine will do, but we prefer hard wood. In our project, we chose white ash, because of its beautiful grain pattern. It is a good idea for this project to use a light colored wood, since that is easier to see and work with in dark.

1) Threaded rod: 1/4 x 20, 1 1/2 inches long, preferably brass.

1) Threaded rod: 1/4 x 28, 6 inches long. This is the drive screw. *(A 1/4 x 20 rod could be substituted here, if necessary, making the tracker about 3 inches longer. Be sure to increase the lengths of two of the wood pieces listed above by about 3 inches.)*

1) Thumbscrew 1/4 x 20, 1 inch long.

3) Tee nuts 1/4 x 20.

1) Piano hinge, 5 1/2 inches long, with mounting screws.

2) Wing nuts 1/4 x 20 thread.

1) Carriage bolt, 1/4 x 20 thread, 3 1/2 inches long.

1) Carriage bolt, 1/4 x 20 thread, 2 inches long.

6) Washers, 1/2 inch diameter

4) Washers, 1 inch diameter.

1) Washer, 1 1/2 inch diameter.

4) Wood screws, 1 inch long.

1) Wood screw, 1 1/4 inch long.

5) Hex nuts, 1/4 x 20.

4) Hex nuts, 1/4 x 28. (If a 1/4 x 20 threaded rod is used, substitute these with 1/4 x 20 nuts.

1) Stiff compression spring, approximately 3 1/2 inches long.

Preparing the Top and Bottom Boards

Construction of the camera tracker is simple. First, cut the sides of the "V", which become the top and bottom of the camera mount. It is best to cut those pieces which need to be glued, so they may set while the other work is being done. Cut two identical pieces 9 1/4 inches long. The wood grain should run the long way for greatest strength. The wood is tapered as shown in the top view plan. The wide end should be 5 3/4 or 6 inches wide; the narrow end should be 4 inches in width. You may want to round the ends of the narrow end of the top and bottom boards for a more professional touch.

Cut another strip of wood, 1 inch wide whose length is the same as the wide end. Glue this piece to the bottom board at the wide end as shown in the side view. (Later the hinge will be screwed to this piece and to the top board.) To glue the wood pieces, use any wood glue, such as Tite Bond, clamp and leave overnight. To avoid marring the wood, always use some wood scraps between the clamps and finished surfaces. Remember when fastening clamps to wooden parts, adjust the clamps to be just "finger tight." Using excessive force on the clamps forces too much glue from the joined surfaces, weakening the bond.

Constructing the Camera Mount

If you have a universal ball joint camera mount, that can be bolted to top piece later, skip to the next section. To make our swivel mount, cut three pieces, each 2 1/2 inches by 2 3/4 inches and glue all three pieces at their faces, making a block 2 1/4 inches thick. One of the long sides of the block will be glued to a 4 inch circular disk; the top side will need to be rounded a great deal. It may be easier to round each of the three pieces before gluing, using a saber saw. After the

glued block has thoroughly set, drill a 1/4 inch hole through the three pieces, centered, 1/2 inch below the top.

Push the 3 1/2-inch carriage bolt through the block and tighten it into place. The best way to lock it into place by recessing the square into the block, is to place a large washer on the threaded end and use a nut and wrench against the washer to slowly draw the bolt head into the block.

From one of the remaining pieces of wood, cut a 4-inch diameter circle. If you have access to a drill press, mount a circle cutter in the chuck. With a circle this size, feed the cutter in slowly. Several times, raise the bit momentarily to allow it to cool, particularly when using hardwood. Without a drill press, this disk may be cut with a sabre saw, but make this disk as circular as possible. A quick way to insure a good circle is to insert and tighten the carriage bolt and mount it in a drill and rotate it against a file or wood rasp.

Photo 1

Next, carefully determine the center of the disk and drill a hole all the way through so the two-inch bolt will make a snug fit. A circle cutter will have provided us with the hole we want. On

Photo 2: An alternate camera mount system

the top board, mark a line three inches to the right from the hinged edge. Slide the circular disk along this line, so the disk edge is 15/16 inch from the closest point on the edge of the top board. Mark the closest point on the edge. Using a scratch awl through the hole, mark and drill a 1/4 inch hole through the top board. Centered on the edge mark, cut a notch 1 1/2 inches wide, 3/4 inch deep.

Before inserting the carriage bolt through the disk, countersink the top so the bolt head is fully recessed and the top is flat. Sand the top of the disk and the large bottom side of the block so they are flat and after centering the block on the

disk, glue them together. A couple wood screws inserted into the block from the bottom of the disk will hold these parts while the glue sets.

Cut two pieces 2 3/4 inches wide, one 3 1/4 inches in length and the other, 4 inches. Make sure the grain runs along the long dimension. On the shorter piece, center and drill a 1/4 inch hole one inch from the bottom short edge. Round the bottom corners and temporarily mount on the carriage bolt to make sure the corners are rounded enough for this piece to swing 180 degrees. In the top, four inch piece, cut a 3/4 inch wide slot, 3/8 inch deep. In the center of the top, drill a 5/16 hole all the way through. Using a 1/2 inch wide flat blade bit or a 1/2 inch Forstner bit, drill a half inch hole into the top, halfway through the thickness. This will leave room for a hex nut and washer to be recessed in final assembly.

Before gluing, you may want to countersink two 3/4 inch flat head screws through the top and screw into the top of the side piece. To do this properly, choose a drill large enough so the screw's shank will easily fit into the wood, and drill all the way through. With a countersink bit, drill the countersink deep enough so the screw heads fill be flat with the top. Using the top piece as a guide, mark and drill corresponding holes into the ends of the side support. Now, glue the top to the side as shown in the side view and tighten the screws.

Assembly

While the glued pieces are setting, return to the bottom board. Three inches from the left (hinged) edge, center and drill a hole all the way through with a 5/16 drill. Following the same hold, drill a half inch hole half way through. Insert a Tee nut, pointing downward into the top of the hole. From the bottom, use a large washer against the surface and inserting a bolt, with a wrench, draw the Tee nut into the wood. This forms the attachment to mount the tracker on a camera tripod.

Place the hinge along the one inch strip already glued to the bottom board and mark the drill holes. Drill the small screw holes and mount

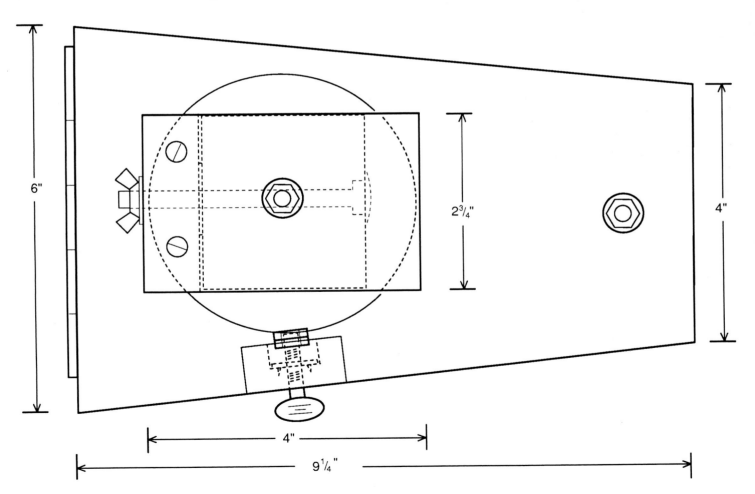

Fig. 1: Camera tracker top view

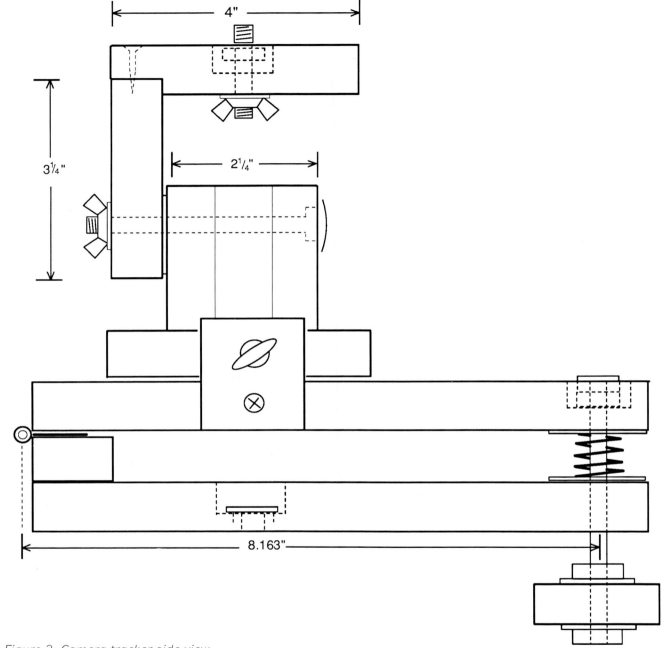

Figure 2: Camera tracker side view

the bottom of the hinge. From the center of the hinge axis, measure 8.16 (about 8 3/16) inches and carefully mark and drill a 1/4 inch hole, centered on along the 4 inch side. (Note: if you are using a 1/4 x 20 threaded rod, the hole will be 11.43 inches from the hinge axis, according to

the formula, above.) Into the top surface of the bottom board, draw in a hex nut, fully recessed into the wood. Use the same technique as with drawing in a Tee nut, working from the bottom side. The bottom board is now finished.

In the top board, drill a 1/4 inch hole in the

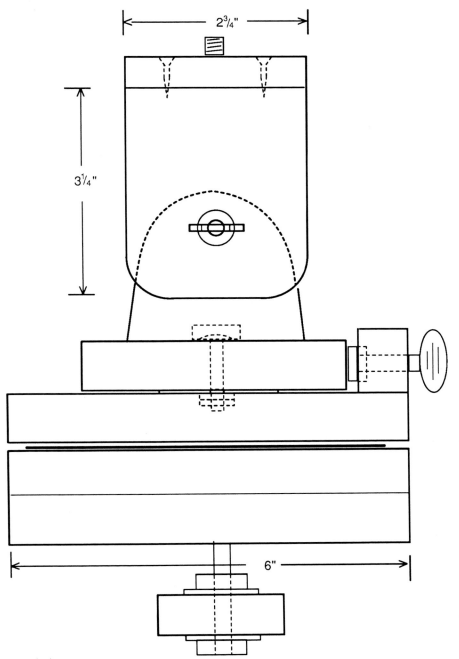

Figure 3: Camera tracker end view

same position relative to the bottom board. Drill this hole with the drill at several angles, to get a slightly elongated hole. Into the top of this hole, center a 1/2 inch drill bit and drill halfway through. Now cut a block 1 3/4 inches long and 1 1/2 inches wide. The grain should run along the longer dimension. This block will fit into the notch in the top piece. Its job is to push a thumb screw against the circular disk of the camera platform to keep it from rotating while an exposure is in progress. Measure 1 1/4 inches from the bottom of the block drill a 5/16 inch hole. From one side drill a half inch hole halfway through. As before insert a Tee nut as indicated in the plans. Now glue the block in place. Drilling and inserting 1 inch wood screw will hold the block in place.

Check out and Finishing

Mount the hinge to the underside of the top board. The mounting screws are probably small and cannot withstand much tightening force, so drill screw holes adequately, especially when using hardwood.

Unless you are mounting a universal ball joint to the top of the camera tracker, follow the steps in this paragraph. Cut a 1 1/2 inch length from the 1/4 x 20 threaded rod. This attaches your camera to the camera platform. On the top end of the rod screw on a hex nut with a half inch washer below it. Leave 1/4 inch of threads above the nut. Put the rod through the top of the camera platform; the hex nut will be recessed in the hole. From the bottom, add a washer and a wing nut. Attach the camera platform to the carriage bolt in the block with a washer on each side and a wing nut on the outside.

Insert a thumbscrew, which will push against the disk, into the block. Lock two hex nuts together on the end of the thumbscrew. To this, you may want to glue some felt or a piece of neoprene or rubber washer, to prevent the thumbscrew from marring the disk when it is tightened.

Place a 1 1/2 inch washer through the bolt

Photo 3: Constellation of Taurus, with Pleiades in lower right

on the bottom of the disk and mount the assembly onto the top. On the underside of top board use a half inch washer and lock two hex nuts together, leaving a little play, so the assembly can rotate freely.

On the long threaded driving rod, attach and lock together two hex nuts at one end. With a washer in the top board, insert the threaded rod, so the locked nuts rest into the recessed hole on top. Push onto the rod between the top and bottom boards, a washer, compression spring and another washer. You will probably have to squeeze the top and bottom together against the spring a bit, but screw the rod into the hex nut in the bottom board. Leave about 2 or 2 1/2 inches of the rod protruding beneath the bottom.

Finally, make a knob to turn the drive rod. Cut a two inch square of wood and make a hole in the center. Round the corners, but make one corner flat. As the screw is turned at one revolution per minute to open the hinge, causing the camera to follow the sky, the flat corner makes a convenient reference point so you can judge how far the screw has rotated. With a pair of hex nuts and washers fasten the handle to the bottom of the drive rod.

Our assembly is finished and ready to try. Try all the motions and make any adjustments necessary. Disassemble, sand all the pieces and varnish. We use polyurethane varnish and typically apply several coats, sometimes with light sanding between coats. After the varnish has thoroughly dried, reassemble. Now you're ready to take some pictures!

With a carefully constructed camera tracker and a little experience, excellent sky pictures can be made, with exposure times of up to ten minutes or so. You may even want to add a one RPM motor to turn the rod, driving the tracker automatically at a uniform rate. The results will improve the more steadily the drive rod is turned and the more accurately the axis of the hinge is aimed at the celestial pole, Polaris.

Technically, this single arm drive will produce an exact drive rate at one instant: the time the rod is perpendicular to the top board. If you are fortunate enough to live away from city lights, where the skies are very dark, you may want to make longer exposures. For exposures of twenty minutes or longer, more sophisticated dual arm drives can be built which give accurate tracking of up to a couple hours. For such examples, look through the issues of *Sky and Telescope* magazine: February, 1988 and April, 1989.

Photo 4: Constellation of Orion (exposure time: 45 seconds)

A Simple 4 1/4 Inch Rich Field Reflector

This telescope making project will provide you with a very useful telescope, regardless of any other telescope you may have. A rich field telescope is a Newtonian-style reflector which has a relative short focal ratio, usually around f/4, permitting wide field of view and low magnification. This is called a rich-field telescope, or RFT, because despite its aperture which limits the faintness of stars which can be seen, the wide field encompasses many stars in the eyepiece. Such telescopes are not for observing planets, but they are terrific for just scanning the Milky Way and looking at remote galaxies. Even though several excellent telescopes are available, our small rich field reflectors still provide many hours of enjoyment.

The telescopes featured in this text are of the Newtonian design. Both Newtonian reflectors and small refractors are simple to design and construct. Successfully building these requires neither years of experience nor elaborate shop tools.

To recap Chapter 1: Minimally, the Newtonian telescope requires:

- a primary mirror with a parabolic curve on its front side which collects and focuses the light;
- a small optically flat diagonal mirror to reflect the light to the side to an eyepiece which magnifies the image;
- a tube which serves to hold those basic elements in a rigid position relative to one another.

The primary mirror rests in a mirror cell which is attached to the tube, and the eyepiece holder or focuser permits small focusing adjustments. Nothing more is needed!

The idea behind this particular telescope is simplicity itself: it embodies an instrument which can be taken outside at a moment's notice, and can be used immediately simply by being cradled in your arms. No mounting, finder aligning, or set up are required! Furthermore, it is fun to build, and entails only a few evenings work to complete.

4 1/4 INCH RICH FIELD TELESCOPE PARTS LIST:

1) 4 1/4 inch f/4 parabolic mirror. If other similar mirrors are used, and the size or focal length varies, all dimensions should be adjusted accordingly and materials of the required size purchased. Measurements included here assume a primary mirror with a 17 inch focal length (with a possible variation of about an inch).

1) Flat diagonal mirror, size 1.25 inches. (A 1.062 inch diagonal will do, but if you use a

wide-field eyepiece, the slightly larger size will provide a more uniformly illuminated field of view.)

1) Tube, at least 20 inches long, five inches inside diameter. We chose a Polyvinyl chloride tube, commonly available as drainage pipe (which has been called affectionately the sewer pipe telescope). Aluminum, fiberglass and rolled paper Sonotubes™ are also good choices. PVC tubing is normally sold in ten foot lengths. Some plumbing shops will sell shorter pieces. Keep an eye out for building construction where pieces of just about the right size may be tossed out as scrap.

1) Standard 1 1/4 inch diameter eyepiece, with a focal length about about 16 to 20 millimeters. (We generally use a University Optics 16mm Konig eyepiece for this telescope, however an Orthoscopic or Plossl also works well.)

1) Low profile focuser with mounting screws (available from several vendors). We will show how to make our own.

1) Length of stiff metal, such as aluminum, about 5 by 1 by 1/16 inch. This is attached to the side of the tube and will support the diagonal mirror.

1) Block of wood 2 by 3/4 by 3/4 inches which will attach to the stiff metal (per illustration), and two #4 by 1/2 inch screws.

1) Tube of Permatex™ flexible adhesive cement, part number 66B.

1) Mirror cell sized for the primary mirror with mounting screws. At the time of this writing, they are available for telescopes of this size from University Optics in Ann Arbor, Michigan. We chose to build our own and full instructions are provided below.

If you choose to make your own mirror cell, the following materials are required:

2) Pieces of hardwood, at least 6 by 6 by 1/2 inch. Wood such as Walnut, Cherry or Maple

finishes and holds up well.

3) Right angle brackets 1 1/4 inches long and three mounting screws. (These are generally used to anchor shelves.)

3) #4 x 1/2 inch round head wood screws and three Nylon washers (Metal washers with felt glued on can be used).

3) Flat head (brass) #10-32 machine screws 1 1/2 inches long, with matching knurled nuts.

3) Round head (brass) #8-32 machine screws, 1 inch long with cap nuts.

3) Compression springs, 1 1/2 inches in length.

6) Small pieces of adhesive felt.

To make an eyepiece focuser, the following are required:

1) Block of hardwood, 2 3/4 by 3 1/2 inches by 1/2 inch.

4) Machine screws #8 by 1 inch with hex nuts.

1) 1 1/4 inch length of thin-walled tubing 1 1/4 inches inside diameter,(preferably brass — for suggestions, see text below.) This is the eyepiece tube.

To make an optional tripod plate for attaching the telescope to a camera tripod, add another block of hardwood and machine screws as for the focuser, and get a 1/2 inch long tee nut with 1/4 by 20 thread. There are two kinds of tee nuts, so use one with small spurs which press into the wood.

Note: For appearance, the screws and nuts should be of the same material. We have employed brass throughout most of our telescopes; however, zinc-plated steel, or stainless steel will do but be consistent.

An assortment of sandpaper sheets in various grades, and a padded sanding block should be kept at hand. Choices of paint and varnish as the final finish are discretionary. For the inside of the PVC tube, a flat black latex paint is necessary; and for the wood parts, apply several coats of polyurethane varnish.

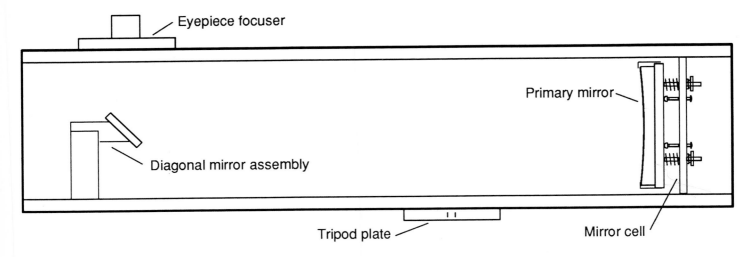

Eyepiece focuser

Primary mirror

Diagonal mirror assembly

Tripod plate

Mirror cell

Figure 1: 4 1/4 inch rich field reflector schematic

Before beginning construction, if you choose to make your own mirror cell, eyepiece focuser and tripod plate with the plans presented here, you may want to photocopy the pages with Figures 2, 3, 5 and 6. To simplify your work later, cut out along the edges of the drawing and use these as templates for marking holes and cut lines.

Preparing the Tube

Tools used in this project include a table saw, hand drill (with a good range of drill bits) and saber or scroll saw. A drill press (or even a small one which accepts a hand drill) will make many tasks easier. It is possible to do this entire project with hand tools; however, greater precision is attained, far more rapidly, and with a greater degree of facility, with a few basic power tools. We have made many telescopes personally with only modest, simple tools. When working with any tools, however, keep safety uppermost in mind. Be absolutely certain that you understand and know how to use power tools correctly!! Remember, too, that when using power saws and drills, eye protection is mandatory.

To greatly simplify working with the tube, a cradle should be built (from scrap wood) so as to support the tube when on the workbench or drill press. Our cradle is rectangular, 24 x 6 1/2 inches x 2 1/2, and is made of 3/4 inch pine stock.

Before cutting ANY material, make sure that the focal length of your mirror is known PRECISELY. Commercially-made mirrors not uncommonly have a focal lengths which vary a few percent from specifications. The telescope tube is pre-

Photo 1

pared, as the first step. It is cut to a length of 20 inches (an inch or so variation will present no problem so long as the error is to a greater length). A rule-of-thumb for these 4 1/4 inch telescopes is that the length of the tube should be the focal length of the mirror, plus 3 inches. For example, a mirror of this size with a focal ratio of f/5 has a focal length of 21 1/4 inches, so the tube should be cut to be at least 24 1/4 inches.

Photo 2

A PVC tube is easily cut with a hand saw, but to accurately square the ends, a table saw is far preferable. Mark your cut line circumferentially around the tube, and cut with several passes on the table saw, rotating the tube carefully each time. A block of wood clamped to the table of the saw as a backstop-guide makes it easier to cut accurately. A large square should be used to check the ends for squareness, then the tube may be sanded or filed as required. Photo 1 shows the ends of the tube being squared using a belt sander.

With a hole cutter mounted in a drill chuck or drill press, cut a 1 3/8 inch hole in the tube, 16 1/2 inches from the mirror end. This operation is shown in Photo 2. Here again, this distance must be scaled relative to the focal length of the mirror. If you have a commercial focuser, drill holes for the mounting screws, using the focuser as a template; then mount the focuser and disregard the next section. Exercise care. Experienced telescope makers tend to snicker a little when they see screws filling misplaced holes. (We've turned perfectly good material into scrap ourselves!)

The Eyepiece Focuser

To see the greatest stars in a RFT, the size of the diagonal mirror (which itself obstructs some of the light) should be kept as small as possible. This suggests a low profile eyepiece focuser, therefore the eyepiece distance from the telescope tube is minimal. This has become a modern trend in telescope-making. A taller rack and pinion type focuser is fine for longer focal-ratio telescopes, but it is not recommended for an RFT. To keep the eyepiece as close to the tube as possible, we make our own focuser for rich-field telescopes of apertures six inches or smaller. This arrangement leaves little room for changing the focus, therefore, the scope is effectively made for a given eyepiece which is always used with that telescope.

A nearly fixed-focus eyepiece holder is simple to make. The hardest task is to find a 1 1/4 inside diameter thin tube, particularly brass. Several vendors once sold it, but we know of no current sources (In our desire to use brass parts, we had to resort to turning a piece of tubing on a

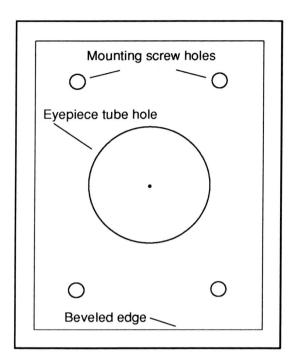

Figure 2: Eyepiece focuser template

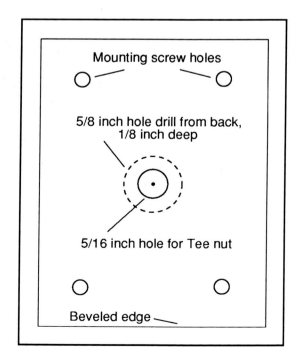

Figure 3: Tripod template

lathe). Occassionally, brass tubing, or chrome plated brass, can be found at a plumbing or hardware store in the form of sink drain pipe (which has a fluted end). If you can find such tubing, cut off and keep the fluted portion. (Aluminum, stainless-steel and PVC tubing, may be easier to find).

Use the template in Figure 2, and cut a block of hardwood 2 3/4 by 3 1/2 inches. Bevel the edges at an angle of 45 degrees. Cut a hole (with a hole cutter), centered in the block which will contain the eyepiece tube. This should insert with a tight fit, so that it can be pressed in with a clamp (protecting the tubing end with a block of scrap wood). Cut two strips of wood in the shape of a wedge, and glue onto the two long edges the back of the block. The thick side of the wedge should run along the outside. After the glue has set, fit this to the radius of the tube, by filing and sanding as necessary. Figure 4 shows the fit of the focuser against the tube. (If you choose to make a tripod plate, while you're at it, prepare two more such wedges for it.)

Drill and countersink four holes for the mounting screws, as shown on the template. Mount this assembly to the telescope, centered on the eyepiece hole. It is not necessary to tighten screws and nuts fully as the parts are mounted in the tube, inasmuch as disassembly for painting will be necessary, anyway. Photo 3 shows two telescope tubes, one with the holes drilled for the eyepiece holder and the other with the focuser mounted without the eyepiece tube inserted.

Adding a Tripod Plate

As a hand-held telescope, this one is ideal. But in order to share your observing enjoyment with others, it is easy to add a plate to the bottom so that the telescope can be mounted on a

Photo 3

camera tripod. (A sturdy tripod is an absolute must!)

As we did with the eyepiece focuser, glue two wedges along the long edge of the back of the tripod plate and fit it against the radius of the tube.

Using the tripod-mount template, Figure 3, drill and countersink the four mounting screw holes. From the back-side, a half-inch spade or Forstner bit is used to drill a 1/8 inch deep counter-bored hole. The tee nut will then fit snugly without a protruding flange. Prior to inserting the tee nut, a 5/16 inch hole, penetrating the board entirely, is drilled using the pilot hole of the counter-bore for centering.

Push the tee nut into the hole. To fit the tee

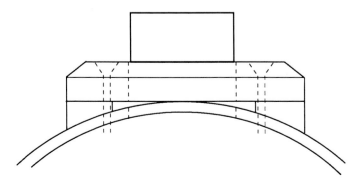

Figure 4: Eyepiece holder mounted on tube

nut without danger of splitting the wood, place a wide washer against the bottom of the plate, and with a nut and bolt screwed fully into tee nut, use a wrench to slowly draw the tee nut into the wood. Mount the tripod on the bottom of the telescope tube, on the side opposite the eyepiece, (which forms the base of the scope) approximately two inches toward the mirror end of the tube, from tube's midpoint.

Mounting the Diagonal Mirror

The metal strip measuring 3 3/4 x 1 inch (called for in the parts list) should be formed as follows: bend a 90 degree "L" from one end measuring 1 inch. The bend must be carefully done, so that the center of the block will center accurately in the circular outline of the tube. Drill two holes in the metal strip (on the short side) for the mounting screws.

Photo 4: The Diagonal Mirror

Cut a 45 degree angle on the end of the 2 x 3/4 x 3/4 inch block; mount the block to one side the stiff metal strip (the angle-side of the block facing upward) with two small screws.

The diagonal mirror is glued to the 45 degree

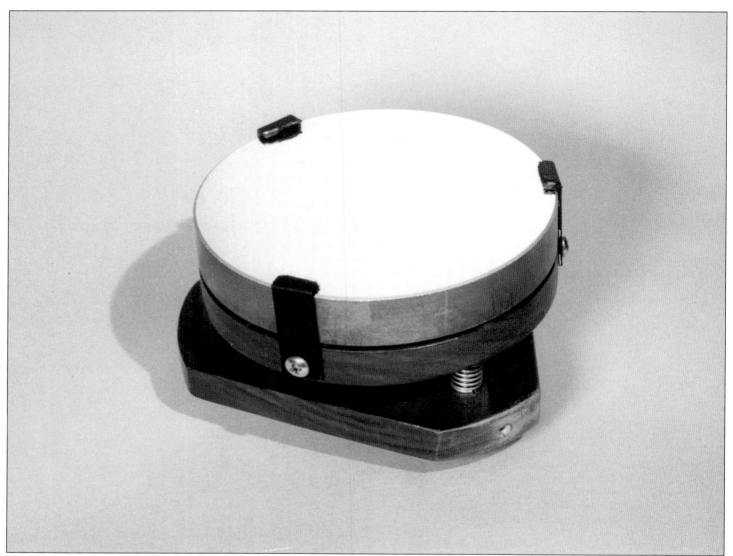

Photo 5: 4 1/4 inch Mirror Cell

face of the block with a flexible adhesive, (such as Permatex) or adhesive tape. The center of the diagonal mirror should be 1/16 inch below the center of the block. For other telescope sizes and focal lengths, a more detailed discussion of how to determine the diagonal mirror placement can be found in Appendix A.

Mount this diagonal mirror assembly to the telescope tube, after the adhesive has set. Insert the assembly into the front of the tube so the diagonal mirror is centered beneath the eyepiece hole. Because the diagonal is not centered on the

block of wood and is therefore not exactly centered in the tube, move your eye toward the eyepiece hole, while holding the assembly in place. Just as your eye approaches the hole, the far end of the tube can seen reflected in the diagonal.

Carefully adjust the position of the diagonal assembly so that the far end appears centered on the outline of the diagonal mirror. Mark that position, measure the marks and transfer them to the outside of the tube.

At those marked positions, two screw holes are drilled from outside; which should be large

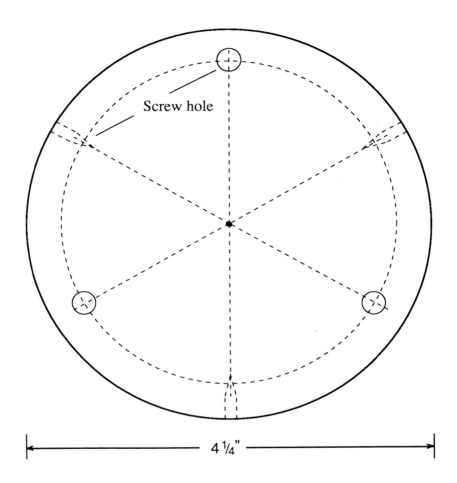

Figure 5: Mirror support disk

enough for the screw threads to pass through. In a PVC tube, countersink the holes so that the flat head screws will be flush with the outside of the tube. Mount the assembly in the tube.

Making the Mirror Mount

A rigid, yet adjustable, mirror mount is essential to maintain optical alignment of your telescope. For most of our small telescopes, we have used a mirror cell of our own design; and which is featured in Project 3. You may want to use it here, but it is a bit fancier and more time consuming to

construct. For this project, however, we will built a simpler cell which meets our needs.

With a hole cutter or saber saw, prepare two disks from half-inch thick hardwood: one 4 1/4 inches diameter (exactly the diameter of the mirror) and the other being the inside diameter of the tube (in this case, 5 inches).

On the smaller disk (which will hold the mirror), choose and mark the side on which the mirror rests. Scribe a circle with a radius of 1 3/4 inches. Around that circle, at 120 degree intervals, mark with a pencil or scratch awl, three

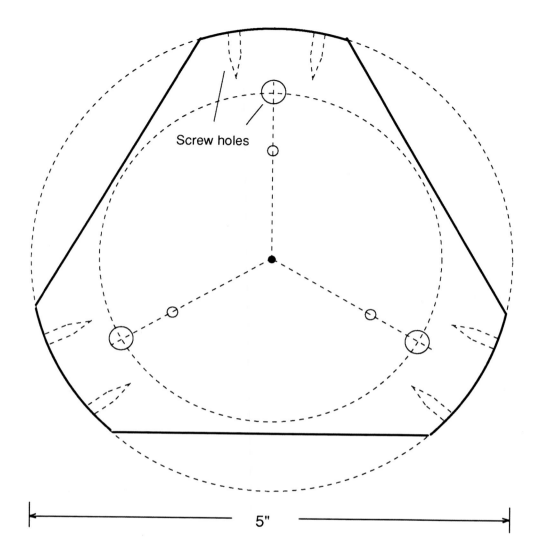

Figure 6: Mirror mount template

points for screw holes to accommodate the three #10-32 x 1 1/2inch flathead machine screws. Make these holes with a drill such that the screws will fit tightly, requiring insertion with a screw driver. For hardwood, a 13/64 inch drill size will be correct. Countersink the holes so that the screw heads are below the surface, because we want the mirror to rest on a flat surface. (Do not put the screws in yet.)

Prepare the right angle brackets by cutting one end to about 1/8 inch. Equally spaced around the edge of the smaller disk, drill three holes for the mounting screws. Offset these holes 60 degrees from the first set. You may need to drill matching holes in the brackets, in which case, drill the holes a little elongated to permit slight adjustments of the brackets. The top of the bracket with the short tab will clip over the edge of the mirror's face.

On the larger disk, you may prefer making three cuts, so the result will resemble an equilateral triangle with rounded corners, as shown in Figure 4. Center the small disk on the larger one; and, using a scratch awl, mark points for drilling

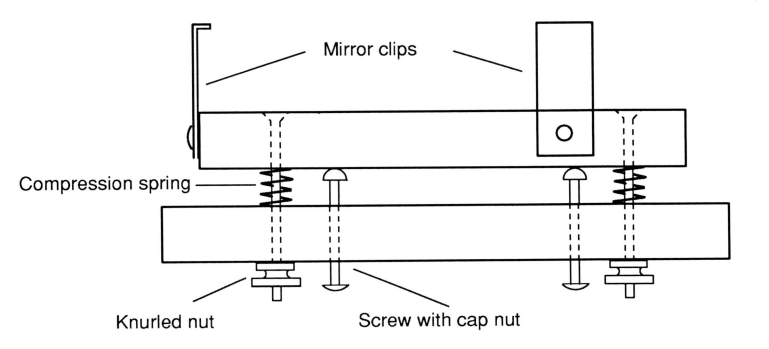

Figure 7: Mirror mount assembly

(through the holes in the smaller disk). Drill these holes to accommodate the #10-32 screws. They should be large enough so that the screws will pass through smoothly. Mark and drill three holes, each 5/8 inch toward the center from these holes, so that the #8-32 inch machine screws fit tightly. This will require a 5/32 inch drill. Screw into the wood (from the side which will be at the rear of the telescope), each of the three #8-32 inch machine screws, deep enough to hold the cap nuts.

We chose to dress up the mirror cell a bit by cutting a 1 3/4 inch circular disk and gluing it to the center of the rear of the mirror mount and then screwing it tight with three countersunk #6 x 3/4 flathead wood screws. This serves to cover the central hole left by the circle cutter. This is entirely optional, but requires only a few minutes to do.

Assemble the mirror mount by putting the three 1 1/2 inch machine screws in from the mirror side of the smaller disk. Place a compression spring on each of these screws. Attach the larger disk to these screws with the knurled nuts,

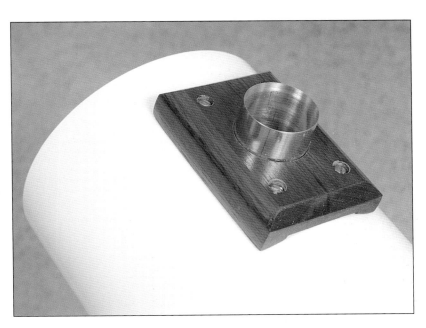

Photo 6: 4 1/4 inch Mirror in Cell

screwed on sufficiently far so the two disks are spaced about a half inch at each point.

Mounting the Mirror

If you are using the mirror mount presented here, or one which has been commerically-made, it is time to mount the main mirror. Cover the mirror surface with a few layers of soft lens tissue and a sheet of cardboard, for protection.

On the mirror cell, upon which the back of the mirror rests, glue three small pieces of felt or rubber to cushion the mirror. Using the three side clips, (made with the right angle brackets) place small pieces of felt between the clip and the mirror's face. Mount the mirror to the cell. Do not fasten the mirror tightly by applying pressure, because temperature changes in the materials may stress the mirror, thus warping its delicate curve and ruining the images. A few thousands of an inch of play is good.

Fitting the mirror and its cell in the telescope tube requires patience. The focuser and diagonal have been installed earlier. The telescope is now lain on a table, or mounted on a camera tripod out-of-doors. The eyepiece is inserted and adjusted to a comfortable position.

The telescope is now aimed toward some distant object at least a few hundred yards away. The mirror and cell are inserted from the rear of the tube. At some point, with the cell an inch or so inside the tube, an image can be seen. By sliding the mirror and cell slowly along the tube, the image of the distant object should come into focus. It will probably not look as good as you might expect, because the optical elements have not yet been aligned. When optimal focus is achieved, mark its position inside the tube without moving the cell.

In the workshop, transfer those marks to the outside of the tube. Rotate the cell so that one of those points is on top of the tube, in line with the eyepiece. This will facilitate final collimation of the mirrors. At each of the three points around the tube, drill and countersink two holes through the tube, approximately one inch apart, so that the screwheads will descend just below the tube's surface when tightened. Cut a section of dowel rod (any size is satisfactory) with a length extending from the front end of the tube the mark which was made at the level of the inside markings minus the thickness of the insert. Then, with the mirror removed from the cell, reinsert the cell and drill through those holes into the cell. The six screws will go through the holes in the tube into the edge of the cell.

Final Assembly

At this juncture, the finish is applied to the wooden parts. We use several coats of polyurethane varnish with light sanding between coats for our scopes. For tips on getting just the right finish, see Allan Saaf's comments in Project 3. The inside of the tube should be painted flat black. A durable paint (such as spray epoxy) that gives a finish which may be cleaned, is chosen for the outside of the tube.

Fit the eyepiece tube to the focuser and install. Add the tripod plate if one was made. Replace the mirror in its cell and install the assembly in the tube. Standing a few feet in front of the telescope (best done with the assistance of a friend) adjust the knurled nuts on the mirror cell so that your reflection is centered the mirror. The optical axis of the mirror must coincide with the long axis of the tube. Once the mirror is aligned, turn each of the three locking screws (with cap nuts) so that the cap nuts gently push against the back of the disk which holds the mirror. This serves as a locking mechanism to keep the mirror fixed in place, while still permitting adjustment later by loosening the screws slightly.

Finally, mount the diagonal mirror. While looking through the eyepiece tube (without an eyepiece), adjust the diagonal mirror by bending its bracket slightly so your eyeball appears centered in the diagonal. This job will require patience and a few attempts at aligning mirrors so that they are exactly right. Small adjustments in the primary mirror may be needed.

Now your telescope is ready for "first light." Take it outside for stargazing. If the star images are perfect points of light, showing no discernible size and good focus you have aligned the optics correctly. Often as not, the "star test" will show less than perfect images. In this case, again check the optical alignment and make the adjustments very carefully. Follow the steps as given at the beginning of Chapter 6. If the images still are imperfect, recheck the primary mirror, to make sure that it is under no pressure from the cell. Also, check the diagonal mirror to ensure that the adhesive is not placing it under strain.

A Hand Held 4 1/4" Rich Field Reflector with a Wooden Tube

Introduction

Many amateurs get about as much enjoyment from building a telescope as from using one. There is certainly pleasure, excitement and fascination in grinding, polishing, and testing optical surfaces, as well as in the design and construction of the mounting. But sometimes amateurs get so involved in these aspects of the project that their first telescope is terrifically cumbersome and almost impossible to transport. Those who live in dark sky areas don't feel they have a problem, of course; they just go out to their observatory building which houses their 32-ton telescope and start viewing. But most of us live in towns and have annoying city lights to contend with; and when we want to enjoy our wonderful new astronomical telescope, we want to go out to some secluded spot in the country where the sky is dark. And people who are lazy like me don't want to spend valuable energy hauling heavy and clumsy telescope parts, or waste half of their observing time putting it together and taking it apart again in the rain.

Obviously portability is an advantage for many amateurs. We want our telescopes to be fairly light and compact, easy to transport, quick to assemble (and disassemble at the end of the session) and to be sturdy. Many telescopes have the first three of these qualities, but are seriously lacking in stability. Some light-weight telescopes are so unstable that star images dance all over the field in the slightest breeze. So the problem is to find a way to combine strength, stability, and compactness with light weight, ease of assembly, and relative simplicity of construction. In the chapters which follow, three telescope designs are described which combine these features. Obviously, there are many different designs which also possess those qualities; but these have the advantage that they are relatively inexpensive, fairly easy to build, and possess wonderful performance characteristics.

The 4 1/4 inch rich field telescopes described here are wonderful for the purpose of deep sky observing of faint objects such as galaxies and nebulae, and I have used mine with real satisfaction for years. A highlight of my experience with it was once observing most of the night on an occasion in which the sky was pitch dark (during new Moon, and the nearest town was 30 miles away) and the sky clarity was perfect. The most remarkable sight was the Cygnus Loop — it was spectacular! It looked like a giant smoke ring, and it seemed nearly to fill the field of view.

This 4 1/4" rich field telescope can either be hand-held or mounted on a photographic tripod or other simple observing stand. Hand-held, it is ideal for scanning the Milky Way, or just roaming around rich areas of sky. But admittedly there are several disadvantages. One is that cradling the telescope in one's arms while viewing makes it impossible to hold it steady enough to present a completely satisfying view. And it's sometimes hard to find things: the observer is looking at right angles to the direction of the object, so it's sometimes awkward to locate an object - it takes a little getting used to. Mounting the telescope on a photographic tripod would provide a steadier observing platform, but it would require constant adjustment in two directions to follow the motion of an object in the sky. Nevertheless, despite the image unsteadiness when using the telescope hand-held, this telescope has been, for me at least, one of the most satisfying instruments I have ever used for viewing faint objects.

And it's beautiful to look at.

Making the Tube

This project requires some woodworking skill and experience, as well as a reasonably well equipped home workshop. In particular, we will assume that the worker has the following home power tools: drill, table saw, saber saw, belt sander, and router. A drill press and a small bench sander are useful, but not essential. A number of hand tools and accessories are needed also:

- a straight router bit (1/2" is good)
- sandpaper - coarse, medium, fine, and extra fine, plus fine steel wool, for hand sanding and final finishing of the varnished tube and accessories.
- sanding drum for power drill (1" - 2" is OK) drill bits, and countersink bit, and an accurate angle measuring gauge or knowledge of geometrical construction (you'll need to make accurate 30- and 60-degree measurements)
- clamps: at least three 6" capacity woodworker's clamps
- large bench vise
- patience

PARTS AND SUPPLIES

I chose walnut because it's beautiful and easy to work with. Some may prefer another hardwood, such as cherry or even a more exotic variety. But some hardwoods are so hard and brittle that they're very difficult to machine. In what follows walnut is assumed to have been chosen.

Obtain 1/2" thick clear walnut. It's available at many specialty woodworkers stores, but many lumber yards don't carry it. Most large towns have such a store. You will need 6 pieces, 4" x 20" long (slightly larger pieces would be better) to make the tube, plus a single piece of walnut about 7 1/2" to 8" wide and 36" long to make other parts of the telescope tube and its mirror cell, eyepiece holder, and mounting bracket. Select pieces of walnut that have pleasing grain, and which have similar color and grain appearance. Be careful the wood is not bowed.

A note on wood thickness: be sure that your walnut is really 1/2" thick. For example, the actual dimensions of what is called a 2"x4" are actually 1 3/4" x 3 1/2".

Supplies checklist:
1) Pint Varnish - semi gloss interior/exterior varnish is ideal (1 pint)
1/2) Pint flat black paint.
1) Can spray lacquer or varnish

35) Flat head brass wood screws size #6 x 3/4 inch

6) Flat head brass wood screws size #8 x 1 inch

3) Each, flat head brass #10-32 x 2" long machine screws and corresponding brass knurled nuts #10-32, plus regular #10-32 nuts

3) Round head brass #8-32 x 1 1/2" long machine screws

3) Cap nuts size #8-32 (to fit above machine screws)

3) Medium tension springs, about 1/4" diameter and 1" long

7) Small washers to fit #4 screws

7) Round head wood screws, #4 x 3/8" long

3) Flat head brass or zinc plated screws size #6 x 3/4"

1) Piece of aluminum or other flat metal stock, 1" x 4" long, about 1/16" thick

1) Tee nut - 1'4"x 20

1) Brass tubing 1 1/4" inside diameter, 1" long

1) Aluminized 4 1/4" primary mirror, focal length 17"

Note: The tube length depends critically upon the primary's focal length. If your primary has a focal length other than 17", the tube length must be adjusted accordingly. Simply obtain walnut panel pieces which have a length 3" greater than the focal length. For example, a 19" focal length would require 4"x22" long pieces of 1/2" walnut, rather than the 20" lengths specified here.

1) Aluminized elliptical diagonal, minor axis 1.25 inches.

1) 1 1/4 inch outside diameter eyepiece, of focal length between about 16 and 24 mm.

Constructing the Hexagonal Tube

The main tube has a hexagonal cross section; the distance between opposite panels is 6" on the outside, so there is a clear 5" hexagon inside the tube in which the optical parts will be mounted.

It is essential that the six walnut panels from which the hexagon is to be made are all exactly the same size, that the angles are cut correctly at 60 degrees, and that the edges of the boards which meet to form the seams are smooth and flat. If this is not achieved, the finished hexagon won't look good: edge seams will gap, and the result will look sloppy. First, set the table saw blade at exactly 30 degrees. If this is your first attempt at cutting and fitting pieces, we strongly suggest you find some 1/2 inch thick scrap wood and make some test cuts and fit the pieces together. A trial wooden hexagon a few

inches long will tell readily how well the angled pieces fit without seams or gaps. Panels for this test can be held together nicely with rubber bands.

The upper edge of the saw blade, viewed by the worker, should incline to the left (i.e., it should point at 11 o'clock). Lock the saw blade angle and height in position.

For each piece, select the worst appearing side to be the inside of the tube, and mark the bad side with a pencil so you won't lose track. When making the cuts on the table saw, this inner surface will always be face down. Set the rip fence very carefully, to the right of and exactly parallel to the blade, and make the first cut a narrow one in order to leave maximum width remaining on the work piece. Repeat this for the other five pieces, so each has only one tapered edge. Next, mark the desired width (3 5/8") on each piece, so the pencil mark is on the side farther from the taper. Set the rip fence, still placed to the right of the blade, so the second cut on the work piece will be just outside the pencil mark, and keep it locked in place for the remaining five cuts. Be sure that the bad side is down for the second cut on each panel. In setting up the blade, you may wish to experiment with a piece of scrap first, to ensure that your expensive walnut doesn't transform itself into scrap. Photo 1 illustrates this procedure. Note: for safety, use a push block for pushing the wood into the saw blade. Note the fingers and the hand position on the rip fence: keeping several fingers on the outside will help avoid accidents.

Photo 1

After the second cut on each panel has been made, you should have 6 identical 20" long panels, which (when viewed from one end with the outside of the panel facing up) should have an upper width of 3 5/8" and a lower width of 3 1/8"; the opposite edges should be tapered in opposite directions. It should look sort of like this: _____/. If not, return to the lumber yard.

Next, reset the table saw blade exactly to 90 degrees, and shave the ends of the panels so the ends are exactly parallel and the panels

themselves are exactly 20" long (for a 17" focal length - see note above.)

The next task is to machine the tapered edges of the panels with a router, to prepare them for gluing. To do this with a router table which can be inclined to the cutter is easy: simply set the angle to 30 degrees and proceed. My router table does not tilt, so I built a small jig for the job. It's simply a thin piece of wood (1/4" x 4" x 12") glued at opposite ends to two V-shaped pieces of scrap. The angle of the V is exactly 30 degrees, so this jig presents the proper angle to the router table surface. Clamp the jig to the router table and adjust the router fences so the cutter would remove only a very small thickness - less than 1/64" on each pass. This arrangement is shown in Photo 2. Next, with the tube panel outside face down, and the router off, make a trial pass to see how much material would be removed. When you're satisfied that only a thin layer will be evenly shaved off at each pass through the router, pass the edges through the router several times on each edge, each time reducing the width very slightly, until the width of the outside surface of each panel is exactly 3 1/2".

Photo 2

Gluing together the panels is done in stages, and to make it easy and accurate, it is easiest to make two small jigs. Each consists of a board about 9" x 20", with two parallel narrow 20" long strips glued on opposite sides of the upper surface. The spacing between the strips must be accurately laid out: on one, the inner edges of the strips must be separated by exactly 6 1/16", and on the other the corresponding separation must be 7".

Photo 3

When two panels are placed in the opening between the strips of the more narrow gluing jig such that two panel edges which will be glued together meet, you should notice that they fit perfectly with no gaps at the joint. Apply glue to the edges to be joined, put a little waxed paper over the seam (so your clamps won't be glued to the work), and apply downward pressure with clamps. Make certain that the corners come together accurately. Too much pressure will damage the wood and force all the glue from the seam; too little may result in a weak bond. Apply just enough force to push the pieces snugly together at the seam, and leave them clamped together overnight, as shown in Photo 3. Next day, repeat this process with another pair of panels.

Next it's time to join a third panel to the pieces already glued up. To do this, use the jig with the larger gap; the procedure is almost the same as before. The third panel, when placed in the jig with the previously glued parts, should make the combination look like one-half of a hexagon, and all the edges should meet accurately at the seams. Apply glue, clamp snugly, and go away until tomorrow. Photo 4 shows the third side clamped after gluing. Note: we found it helps make a better fit by first making two blocks of scrap wood, meeting a an angle of 120 degrees. In the situation shown here, the wood was bowed slightly, and this technique insured a good fit. (Note: applying pressure with strong rubber straps may be a suitable alternative to woodworking clamps.) Next day, repeat this step to make the other half of the hexagon.

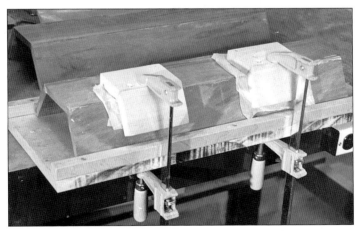

Photo 4

Finally, join the two halves to make the completed hexagon. Clamps, a bench vise, or rubber straps may be used to apply gentle pressure while the glue sets, as in Photo 5. After removing the tube from the clamps, carefully chip off excess glue. Sand the ends of the tube with a bench sander if you have one; the ends should be as even as you can make them.

Photo 5

Next it's time to make the end caps. At the front of the telescope is a hexagonal piece of walnut with a 5" hole centered on it; the back has a 4" diameter hole. If you cut the holes with a saber saw, it's advisable to make the hole before cutting the hexagonal outline. Whether you use a drill press and a circle cutter or a saber saw, cut the hole slightly undersized, smooth it with a sanding drum in a hand drill, and finish it off with hand sanding. After you're happy with the hole you just made, center the end caps on the telescope, mark all the parts so you know their exact final position, and cut the hexagon (but make it a little too large: excess will be sanded off later) with the table saw.

If you have a drill press and a circle cutting tool, it is easy to cut the hexagon first, find the center and clamp the work to the drill press and slowly lower the cutting tool into the wood. Photo 6 shows the beginning of the hole cutting operation. Make sure there is adequate clearance all the way around the circle for the cutting tool to avoid hitting the clamps when it is deep into the wood. The completed cut is shown in Photo 7 before the center is lifted out.

You now have the two end caps, each with a hexagonal shape (but a bit too large), ready to

attach to the telescope tube. The front cap will be glued permanently in place, but the back must be removable so you can later install the mirror cell. Mark the sites for the mounting screws on the front and back end caps (each mounts to the tube with twelve #6 flat head wood screws, 3/4" long), drill and countersink the holes, and attach the end caps with zinc plated screws the same size (these screws are just temporary.) Glue on the front cap now, but be sure not to glue the back.

With end caps attached, the tube is ready to be given its final shape. This is best done with a hand held belt sander while the tube is held in a large woodworker's vise. Be careful to remove material gradually and evenly. As you work, any the excess material on the end caps will be removed also, assuring a perfect fit. You will probably want to round all the corners of the tube along the seams, too; this is easily done with the belt sander. As soon as you're happy with the shape, turn off the sander and put it away. Final shaping and smoothing should be done by hand with a padded sanding block, to assure better control of the final shape. Use medium grit paper, and finish off with fine and then with extra fine grits. As always when sanding, be sure to sand with the grain. Smooth the tube carefully and thoroughly, to remove scratches left from the coarser sandpaper.

Remove the screws from the end caps and remove the back cap. (Mark the back so you can remount it in the same place later!) It's time now to drill the eyepiece hole; that hole should be centered in one of the panels, 3 1/2" from the front of the tube, and should be 1 1/4" in diameter. A hole saw in a hand drill or a drill press is a good tool for this purpose; spade bits make a very jagged hole and should never be used in fine woodworking.

Photo 6

Photo 7

The eyepiece holder and the mounting bracket come next. Each is a rectangular piece of 1/2" walnut, 2 1/2" x 3 1/2". In one, drill a 1 3/16" hole at the center with a hole saw if you have one this size; otherwise, drill it to 1 1/4" and file or sand it larger. The size of this hole must be just right: the eyepiece tube fits inside, and you want your eyepiece to be smoothly adjustable. The mounting bracket has a 1/4" tee nut mounted at its center. Each of these pieces should have four mounting holes drilled and countersunk near the corners. They will attach to the tube with brass #6 x 3/4" screws later on, after all the parts have been varnished and given their final sanding and hand rubbing.

Making the Mirror Cell

The primary mirror must be mounted securely in place, its alignment must be adjustable, and it must be held in its cell gently to avoid strains which could cause it to change the shape of its reflecting surface. These requirements are all met in the mirror cell described here.

The mirror itself is held in place on a 4 1/4" diameter walnut disk. At three evenly spaced intervals around the edge of the disk are attached small wooden cleats which extend to the mirror surface. Into the ends of the cleats are inserted three #4 round head wood screws, 1/2" long, which hold down washers hat press gently against the mirror surface. These washers

103

hold the mirror in its cell. Pads between the washer and the mirror surface may be used to avoid scratches.

First, make the 4 1/4" walnut disk and put it aside until later. (This could be made from the 5 inch hole cut from the front end cap.) Next, make the three support legs which will hold the cell in place within the tube. Starting with a piece of walnut 1 3/4" wide and about 9" long, mark a point along the midline, 2 1/2" from one end. With extreme care, cut (a miter box is good if you have one) through the point you marked, with the saw set at 30 degrees. Next, from the small piece you just cut, remove the sharp end of the piece, by again cutting at 30 degrees through the point marked on the work piece. If this is done correctly, you will have a piece 2 1/2" long and 1 3/4" wide, with one end straight and the other end forming a triangle of 120 degrees. It should look like: ⬅️. In a similar way, make two more legs from the stock material.

Photo 8

When the three legs are put together so their angled ends meet, you should have a piece shaped somewhat like a 'Y' in which each leg of the 'Y' is 2 1/2" long; the pieces should meet precisely at the center. Glue these ends together by setting them on a flat surface and applying pressure. You may wish to make a small jig from scrap material so you can apply pressure evenly and make a better joint. Photo 8 shows two such mirror cells glued in their jig.

Next, use a circle cutter to make a walnut circle 1 1/2" in diameter. Sand it smooth and glue it to the center of the 'Y', in order to reinforce the joint at which the legs meet. You may wish to strengthen the 'Y' still further by drilling and countersinking three holes in the circle to accommodate wood screws extending into each leg.

Now it's time to fit the parts together. In order to get the holes to match exactly in the pieces to be joined, drill and countersink a hole in the center of the 4 1/4" circle and screw that disk (just temporarily) to the 'Y', centering it exactly on the side opposite to which the small circle is attached. Mark the location for the three mounting screws (the 2" long flat head screws) and drill holes of the proper size for them through both pieces of wood. The holes should be placed 1 1/2" from the center, along the midline of each leg of the 'Y'. Next, countersink the holes in the mirror mount circle to accommodate the mounting screws. Finally, separate the large circle from the 'Y' by removing the screw; make sure to mark the proper orientation so you can put it back again the same way.

Enlarge the holes in the legs of the 'Y' so the mounting screws move easily in them without binding. Next, drill and tap three more holes in the legs of the 'Y', to receive the #8-32 brass machine screws. Be sure to drill the holes small enough the screws fit tightly, requiring a screw driver to insert them. (Alternatively, use tee nuts instead of tapping the hole in the wood.) These holes should be located 1 1/8" from the center of the 'Y'. Finally, put the mounting screws through the countersunk holes in the 4 1/4" circle, and secure them with nuts on the other side.

Set the mirror on the mounting circle and make three small wooden cleats, each about 1/4" thick and 1/2" wide, which attach (with wood screws) to the rim of the circle and extend to the mirror surface. It's a good idea to round off the cleats with a file, to fit the curved edge of the circle. Drill small holes in the tops of the cleats, to fit the 1/2" long #4 round head wood screws which will attach the washers that hold the mirror in place. In completing this phase of the project, it is important that the cleats or the washers not apply any significant force to the mirror. The cleats and the washers should hold the mirror snugly in place, but even with the washers tightened you should be able to turn the mirror in its cell. If you can't, it's too tight. And it's a good idea to use felt or cork pads between the washer and the mirror surface to avoid causing scratches. Test the tightness now, before the parts are varnished, so you can file off extra material from the cleats. Once the fit is just right, glue and screw the cleats to the mirror circle.

Finally, take the 'Y' of the mirror cell and see if it slips smoothly into the back of the tube. It should move freely. File off excess material from the 'Y' if needed to make a smooth slip fit.

Making the Diagonal Mirror Holder

The diagonal mirror will eventually be glued to a small piece of wood using a flexible adhesive such as Permatex™. First, take a piece of wood 3/4" x 3/4" x 3 1/2" long, and cut one end to taper exactly 45 degrees. Next, take a piece of aluminum or other metal 1" x 3 1/2" long, about 1/16" thick, and bend it in a metal vise into an 'L' shape, in which the short leg of the 'L' is 3/4" long. Drill two holes at the center of the short leg of the 'L' for short #4 mounting screws; these will eventually attach the diagonal holder inside the tube. Attach the straight end of the wooden piece to the top end of the 'L' with short screws. The distance from the bottom of the 'L' to the center of the wooden square should be exactly 2 1/2": no more, no less. When this part is assembled properly, you should be able to place the short leg of the 'L' inside the tube near the front; and looking into the eyepiece tube, you should look directly down upon the 45 degree taper to which the diagonal mirror will be attached. After you're satisfied that this looks right, paint the inside of the tube and the diagonal holder flat black.

Finishing and Assembling the Tube and its Accessories

Varnishing has not been mentioned yet, because nothing should be varnished until after all the parts have been constructed and given final sanding. In fact, it's best to do the final finish on all the parts before attaching them to the tube: that includes the eyepiece holder, mounting bracket, and the telescope back. The mirror cell should not be assembled yet either, although the three long screws on the mirror circle can be left on during varnishing.

None of the other brass screws should be attached yet either. Brass left in air tarnishes, of course, and to give the finished telescope a really professional touch, all the screw heads should be varnished also. Good results can be obtained by first sanding the heads on very fine (#400) grit sandpaper until they are shiny, and spraying them with varnish. Make a rack from a piece of scrap wood and drill enough holes in it to hold all the screws you will need for the project.

Spray them all at once, and put the rack in a 200 degree oven for about 15 minutes. This will give the screws a tough finish and keep them looking like new for many years.

Before applying varnish to any of the wooden parts, it's necessary to find the proper location for the mirror cell in the tube. To do this, it's necessary to assemble everything to the tube - mirror cell, diagonal holder, and the eyepiece holder. (Suggestion: use ordinary zinc plated screws, not your fancy brass ones, for this. Save the nice ones for later.) Put the mirror in its cell, fasten down the cleats so the mirror doesn't fall out, and attach the diagonal mirror to its wooden support with Permatex™ or other flexible adhesive such as silicone. But don't take this step until you have read and believed the section of the book which describes proper placement of the diagonal mirror! The center of the diagonal mirror's surface should not be at the optical center. Accurately placing the diagonal is critical to proper alignment. First, place a small indelible mark on the diagonal mirror's surface at the point which should intercept the optical axis (see again the diagonal placement section for a diagram: the distance of the offset is just about 0.09" for a 4 1/4" f/4 mirror.) Apply the adhesive, position the diagonal correctly, and let the glue set up. Then place the finished piece in the tube; when the diagonal is correctly positioned, you should see that mark lined up at the center of the eyepiece hole as you look straight in; hold your head back about a foot to give you a good perspective. When you're satisfied, mark the holes and attach the diagonal support inside the tube with #4 x 3/8" round head wood screws.

Put the eyepiece in the holder at the spot you want it to rest while you are observing. It's good to place it as close to the tube as possible—with, say, about 1/2" of it inside the holder. Now take the telescope tube and the mirror cell with the mirror mounted in it out in the yard, and set it on the picnic table. Aim it at some distant tree branch, at least 100 yards away, slide the mirror cell up inside the back end of the tube, and move it around in there until the branch is in focus. Never mind that the image is lousy: you haven't aligned the mirror yet. When you're satisfied that the image is as clear as you can get it, mark the position of the mirror cell legs inside the tube with a pencil, and go back in the house before you get arrested as a Peeping Tom. Using the marks you made on the inside of the tube as a guide, carefully mark, drill, and countersink six holes in the tube such that two #8 x 1" flat head screws will enter each leg of the mirror cell's 'Y', evenly and symmetrically. This can be difficult, and it requires precise measurement and patience.

Now that all the holes are drilled, and everything fits right, take everything apart and get ready to varnish the tube and the other wooden parts - but varnish each of them individually, and assemble them later.

There many experts giving advice about how to apply varnish, so I'll join the assembly. Here's what to do:

How to Varnish Stuff

1 Thin the varnish with about a tablespoon of paint thinner (not lacquer thinner!) per pint of varnish, and stir it awhile. Shake it too, if you want to; that business about not shaking varnish to avoid bubbles is nonsense. There will be bubbles anyway, but thinning the varnish causes the bubbles to burst before the varnish dries.

2 Take a piece of clean flannel rag and moisten it slightly with linseed oil to make

what is called a "tack rag". The cloth feels a bit sticky, and is used to wipe dust from the piece about to be varnished.

3 You have already made sturdy wire hooks (pieces of coat hanger are good) to suspend at a convenient height each of the parts to be varnished. Make sure the one for the tube is strong enough: on the first walnut tube I built, the hook failed and the tube shattered on a concrete floor.

4 Work fast. Varnish dries quickly, and thinned varnish dries quicker. To avoid leaving brush marks, apply the varnish very quickly indeed, and don't go over wet areas with the brush. Drips will accumulate at the bottom, and can be gently removed after the work is hung up on its hook.

5 Let the work dry over night.

Next day, using very fine sandpaper, sand all the varnished parts lightly, using a padded sanding block. Starting with step 2, repeat the process at least twice more, until at least three thin coats have been applied. After the application of the last coat, and a final very light sanding, smooth the work with ultra fine steel wool, apply furniture wax, and buff with a soft cloth. You're done with the varnishing!

Final Assembly

Now you're ready to put it all together. It may be necessary to touch up the inside of the tube with flat black paint, in case some varnish has dripped inside; this would be the time to do that. Next, attach the wooden parts: eyepiece holder and mounting bracket, being careful to smooth the backs of these parts if varnish has dripped there. Fit the brass tube into the eyepiece holder, making sure that the eyepiece can move freely. Mount the diagonal mirror holder, using the screw holes made when the telescope was assembled earlier. Assemble the mirror cell, being sure to put the #8-32 screws into the legs of the 'Y' and attaching the cap nuts. The springs fit between the disk and the 'Y', on the long #10-32 machine screw shafts. Finally, the knurled nuts hold the two parts of the cell together.

Put the mirror into its mounting, and tighten the washers to hold the mirror in place. Slide the mirror cell into the tube and install the 1" x #8 flat head screws which hold the cell in the tube. Finally, after you put in the screws which hold on the back and which help decorate the front, you can say that you're done.

All except for collimation, or aligning the optics, that is.

Joann Kalemkiewicz, a telescope maker from the Capital Area Astronomy Club, shows the completed telescope being used hand-held.

A Six Inch f/3.3 Equatorially Mounted Reflector

Introduction

After having completed and enjoyed using the rich field telescope described in the previous chapter, I thought: if the views in the 4 1/4 inch f/4 were spectacular, wouldn't they be even more dazzling with a larger aperture telescope mounted on a stable tripod? From this errant thought was born the compulsion to build a very short focal length 6 inch telescope with an appropriate equatorial mounting. I settled on a focal length of about 20 inches (an f/3.3 objective mirror), which would provide the feature that nebulae viewed through it would have the same surface brightness as they would if observed through the 200 inch Hale Telescope on Mount Palomar. But this one would be portable.

(Note: the Hale Telescope, of course, can see many more stars than our 6 inch. Stars appear as "point light sources," i.e. they are so far away, they appear to have no size, regardless of the magnification used. The brightness of stars which can be observed with a telescope depends upon the telescope's aperture. However, the story is different for extended objects such as galaxies and nebulae. For these objects, the apparent brightness depends upon the focal ratio, rather than aperture).

Making this particular mirror was a spectacularly difficult piece of work, and I would only recommend it to an enemy. If I were to do it over again, I would make an f/4 mirror, which would be far easier to parabolize; this mirror took almost forever to achieve a proper figure (at times I thought I should be using a chisel to remove glass rather than polishing on a pitch lap) because an f/3.3 parabola is very steep. Surprisingly, parabolizing an f/3.3 mirror requires polishing away a volume of glass 78% larger than for an f/4 mirror of the same aperture! Take my advice: if you construct a telescope like this one, adapt the design to accommodate an f/4 mirror instead.

PARTS LIST FOR THE 6" TELESCOPE

Telescope tube and accessories:

1) Sonotube™ (or equivalent) 7" inside diameter and about 24" long (depending upon focal length of the primary mirror).

1) Mirror Cell (build or purchase: the one described here is modeled exactly after the cell described in detail for the 4 1/4" walnut telescopes).

1) Diagonal Mirror and Support. The diagonal mirror should have a minor axis of about 1 3/8.

6) Flat head brass wood screws, #8 x 1", to

mount the mirror cell.

16) Round head brass #10-32 machine screws, 1 inch long, and matching nuts.

2) 1/2" Bolts, 3" long.

Hardwood (cherry in this case, but maple or walnut will work well) to make two pieces 2 1/2" x 4" long x 1" thick (to be cut later).These will make the eyepiece holder and optional tripod plate.

Hardwood to make two 3" diameter wooden disks, described in the text.

1) 1 1/4" inside diameter by 1 inch long brass tube for an eyepiece holder. If brass is unavailable, other material such as PVC or aluminum could be substituted.

Mounting yoke and tripod parts:

1) Dowel, 1" diameter, 8" long.

3) Pieces 3/4" plywood, 14" square, glued together to make about 2 1/4"thickness, to be cut into 'U' shape for the yoke.

3) Pieces 3/4" plywood x 12" square, glued together to make about 2 1/4" thickness, to make tripod head.

4) Pieces 2"x4" pine, 6" long, glued together to

make about 6" thickness. This will be shaped and drilled with a 1" hole to accept the polar axle dowel.

6) Pieces 1" x 4" pine, approx. 30" long, to make tripod legs.

3) Pieces 2" x 4" pine, approx. 6" long, to attach to tripod head, to support the tripod legs.

3) 3/8" Carriage bolts 4" long, to mount tripod legs.

3) 3/8" Nuts.

6) 3/8" Flat washers.

6) 1/4" x 4" Carriage bolts to attach tripod leg supports to the tripod head.

2) 3/8" x 4" Lag screws to attach polar axle support to tripod head.

9) Pieces 2" x 4" pine approx. 5" long, to make tripod leg spacers.

18) Flat head wood screws, #10, 1 1/2" long.

6) Flat head wood screws, #10, 3 1/2" long.

Paint: the inside of the tube should be flat black. Latex paint is suggested because of the extremely dull texture.

Paint or varnish for other wood parts if desired. Wood glue should be used whenever joining wood parts.

Making the Tube

After investing an inordinate number of hours making the mirror, it was time to construct the rest of the telescope. In an effort to minimize weight as much as possible, I chose to make the tube of a rolled paper material called Sonotube™. Its light weight and strength make it an excellent choice for telescope tubes of almost any size, and the material is very easy to drill and cut. This tube is 7" in diameter and 25" long. If you choose to build this telescope with a longer focal length mirror, be sure to purchase a longer tube.

I purchased a diagonal holder and an elliptical diagonal mirror (see Figure 4) from Kenneth Novak, another supplier of high quality machined parts for amateur telescopes. The remaining parts can be built easily with the instructions and tips presented here. The mirror cell, made of walnut, is shown in Photos 1 and 2. It is constructed along the same lines as the 4 1/4" mirror cells described in the preceding chapter, with altered dimensions to accommodate the larger mirror size. A commercially-made aluminum mirror cell may be used rather than the one presented here.

Rather than purchase a rack and pinion focuser, I decided to use a simpler design. In Photo 4, the eyepiece holder (on the left) is made from a piece of 1" thick cherry, with a 1 3/16" hole drilled in it to accept the brass tube which holds the eyepiece in a slip fit. The observer adjusts the focus by moving the eyepiece in or out. This focuser is similar to the ones used in the two previous projects.

If you decide to purchase a focuser, especially for a richest-field telescope such as this one, be sure to use one which has a low-profile design, which keeps the eyepiece as close to the tube as possible. This will in turn provide the widest possible field of view without requiring an extraordinarily large secondary diagonal mirror. The newer helical focusers adjust the focus by turning a threaded eyepiece tube which raises and lowers the eyepiece. A rack and pinion focuser is fine for longer focal length telescope where you may want to use a wider range of eyepieces, but they aren't recommended here.

The lower surface of the wooden eyepiece holder is curved to match the tube's outer radius —- in fact, several other wooden parts which attach to the tube are similarly machined. This makes for a nice fit, and it's easy to do with a table saw and a dado cutter. Use a dado cutter set which has the same diameter as your tube, and adjust the height of the blade above the table surface carefully to shave off the right amount from the bottom of the workpiece. The desired part should made from a considerably larger workpiece; the part will be cut from the end of the workpiece later. Clamp a straight edge guide on the table saw surface to hold the work in place as it is fed in a direction perpendicular to the usual cutting direction, with the center of the workpiece directly above the cutter's center. This can be a very dangerous opera-

tion on a table saw, so you should take all the proper precautions and exercise great care. If a table saw and a dado of the correct diameter is unavailable, the tube's curvature can be accommodated to a flat surface by attaching shims of the same material and sanding them to fit. Alternatively, the whole job may be done by hand with a rasp.

Located 6 inches from the front end of the tube, drill or cut a 1 1/4 inch diameter hole for the eyepiece. Mount the eyepiece focuser on the tube, centered carefully on the hole. Next mount the diagonal mirror in the holder. The diagonal mirror and its adjustable holder is held in place by four vanes. The whole assembly is called a "spider." Locate the spider in the tube, so the diagonal mirror is roughly centered when peering in the focuser. Mark the four locations for the mounting screws and drill the mounting holes and mount the spider assembly.

Now that the focuser and diagonal are installed, mount the primary mirror in its cell. Using our homemade cell, between the back of the mirror and the surface on which it is resting, glue onto the wood, three pieces of felt, rubber or cork to cushion the mirror by allowing for changes in temperature. Pieces of felt should be used between the restraining clips or washers and the surface of the mirror. When working with the mirror in its cell, be sure to cover the delicate reflective optical surface by covering it with soft lens tissue and cardboard.

Take the telescope outside and aim it at something a few hundred yards away. Make sure the diagonal is positioned properly and adjust it, if needed. When looking squarely down the eyepiece tube (without an eyepiece), you should see a centered reflection of the rear of the telescope. Now, insert the eyepiece and position the eyepiece where is it most comfortable for observing. From the rear of the tube, insert the mirror and its cell, sliding it slowly along the tube until you see a focused image through the eyepiece. Mark on the inside of the tube the position of the cell. Transfer those marks to the outside of the tube and drill mounting holes then mount the mirror and cell in the tube.

Making the Mounting

In keeping with the goal of minimizing weight, it was decided the usual German-style equatorial mounting with its counterweight, would be unacceptable. A simple fork mounting design, which uses no counterweight was chosen instead. In this design, the polar axis is simply a wooden dowel 1" in diameter and 5" long. When constructing the tripod head, it's better to cut and glue together the pieces of 2"x 4" first, then drill the hole for the polar axis dowel using a spade bit. If you try to laminate pre-drilled

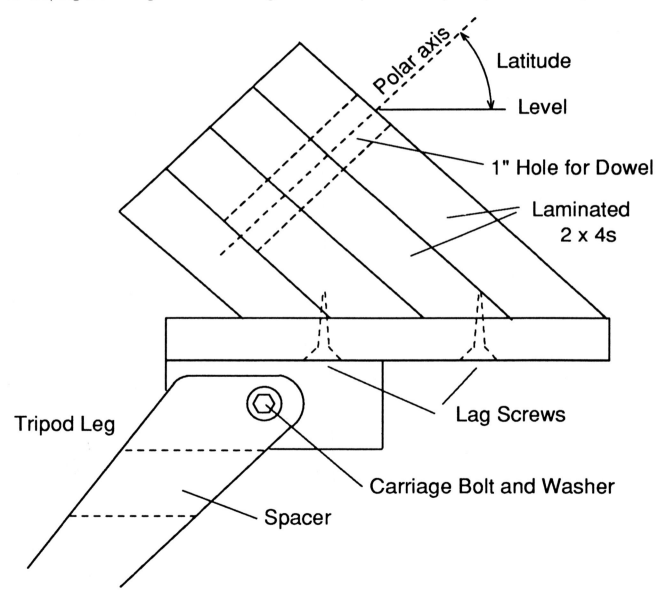

Tripod head assembly

pieces of 2"x 4", the holes won't match up exactly and the dowel won't rotate smoothly. (When one end of the dowel is inserted in the hole, the dowel becomes the polar axis.) Finally, cut the bottom at the proper angle to accommodate the observer's latitude. This angle must be cut so the angle between the polar axis and the horizontal plane is the same as your latitude. Figure 1 shows the tripod head assembly with one leg attached, and illustrates the correct angle for the polar axis.

Each tripod leg consist of two side pieces and three spacers made of 2" x 4" blocks. The spacer at the leg's top is important for its placement. The top of this spacer block rests against the 2" x 4" section through which the lag screw is inserted. This permits each leg to extend to 30 degrees from the vertical. The weight of the mounting pushes against this block. This feature makes the mounting extremely rigid. After the tripod legs have been constructed, attach them to the tripod head with carriage bolts.

The other end of the polar axis dowel is driven into a hole drilled at the bottom of the fork, which is made from a 'U'-shaped piece cut from three layers of 3/4" plywood. These plywood pieces were glued together to make a strong laminated material nearly 3" thick from which the fork was cut with a saber saw. Photos 6 and 7 show the yoke and the polar axis. Care must be taken to cut the inside width of the 'U' accurately, because the tube with its mounting disks must slip smoothly into that space and be free to move easily without wobbling.

In a yoke mounting, the telescope tube is placed between the tops of the 'U' of the fork. In the design described here, the tube swings on bolts which pass through the top of the yoke and are threaded into wooden disks attached to the tube. The axis on which these bolts turn is the declination axis, which provides the other essential axis of rotation for an equatorial mounting.

To achieve telescope balance, it is necessary that the declination axis pass through the center of gravity of the finished telescope tube. The tube must be assembled, with mirror and other optical parts in place, to locate the balance point accurately.

Photo 5 shows these disks which attach to the tube. They have curved undersides, like the eyepiece holder described above, to allow them to fit neatly on the tube's surface. These disks must mount exactly opposite each other, so their outer surfaces are parallel planes. The planes are separated by a distance equal to inside width of the yoke. Thus, you should make the disks first, attach them to the tube, measure their separation, and then cut the yoke so that the telescope fits properly.

The declination axis is formed by a pair of 1/2" diameter bolts on opposite sides of the yoke, which pass through the yoke and into the mounting disks described above. It is necessary either to thread the disks to accept these bolts, or to fit the disks with tee nuts; otherwise the whole telescope would fall apart. Threading wood is always difficult and awkward, and is usually thoroughly unrewarding. Woodworkers understand: wood — especially certain hardwood — wants to split. Most woodworkers know this and accept it; at best it is frustrating and annoying, at worst it is infuriating, and a practioner sometimes regrets not having taken up stamp collecting instead.

Nevertheless, when threading wood works (it sometimes does), one feels enormous satisfaction. Having made every conceivable mistake about this, I feel qualified to offer expert guidance. Here are some suggestions:

1. Make several extra pieces from which the part may be eventually be formed. You'll need them.
2. Drill a hole to be threaded into the piece of wood on a drill press, with care given to choosing a drill size equal to the inner diameter of the screw threads.
3. Get hold of a tap and a tap wrench of the right size for the bolt which is to be threaded.
4. Clamp the piece you intend to thread so the grain of the wood is compressed by the clamp jaws — that is, the clamp jaws are parallel to the wood grain. It will probably split anyway, but it's interesting and instructive to observe how every known law of nature fails when it comes to woodworking.
5. Grasp the clamped workpiece in your left hand, and the tap in your right, and try to envision what will go wrong next. Learn to live with the fact that your imagination failed you.
6. Accept the fact that your first piece of material will probably be spoiled, and try the next. (Armed with this knowledge, I have occasionally thrown away perfectly good, undamaged pieces of wood, because I knew they would be spoiled the first time a tool touched them. I recommend this as a useful labor saving strategy.)
7. Try to thread the workpiece, remembering to thrust the tool downward while performing clockwise twists; oiling the cut while performing a threading motion removes waste material when you make counterclockwise twists of the threading tool.

Anyone daunted by this description of threading wood may want to use tee nuts instead.

This short focal length, wide-field, telescope has no finder — it has always seemed unnecessary. However, a simple lensless finder, essentially a pair of pointers made with eye screws could be added.

Collimation

Once all components have been made and assembled and checked for accurate fit, the telescope can be disassembled for final finishing and painting.

To align the optics, once the mirror is securely in its cell, stand in front of the telescope looking down the center of the tube at the mirror. Using the adjusting screws on the mirror cell, have a friend behind the telescope adjust the tilt of the mirror so your reflection is carefully centered. We want the optical axis of the mirror to be parallel with the long axis of the tube.

Now look at the diagonal mirror through the focuser (without an eyepiece) using its adjustment screws align the mirror so the reflection of your eye (reflected from the primary mirror and the diagonal) is centered in the focuser and in the diagonal.

Photo 1: The 6" f/3.3 mirror mounted in its walnut and maple cell. The parts and the design for the cell are exactly the same as for the 4 1/4" wooden cell described in the preceding chapter. Note that the three support legs are rounded to fit neatly within the telescope tube.

Photo 2: Close-up of the 6" mirror cell, showing additional details.

Photo 3: Diagonal mirror and its holder, made by Kenneth Novak. The diagonal mirror is held against a raised lip by wadded cotton inside the tube. This is an excellent example of the high quality amateur telescope parts available commercially. Notice the small dot near the center of the diagonal mirror. It marks the point at which the diagonal mirror must intercept the optical axis, and is offset from the diagonal's center by 0.14". See the note in Appendix A on diagonal placement for details.

Photo 4: Left: Eyepiece holder with 1 1/4" brass tube and an eyepiece. This simple design, without rack and pinion adjustability, is very effective for short focal lengths and low magnifications.
Right: Mounting bracket with 1/4" tee nut at the center to permit mounting the telescope tube directly to a standard photographic tripod. Note that the undersides of these parts, as well as the disks shown in Photo 5, are curved to fit the telescope tube's outer surface. The holes for the mounting screws are drilled at a small angle so they will be perpendicular to the tube.
Providing the undersides of these wooden parts with a properly curved surface to fit the tube may be done by hand with a wood rasp, or with a dado on a table saw. Better results can be produced with the table saw, but it is tricky and can be very dangerous unless you are experienced and careful. Here's how I did it: I used a table saw blade of the same diameter as the telescope tube, and raised the blade so it cleared the table surface by a distance equal to the depth of the desired cut. A piece of straight wood should be clamped to the table saw top, and positioned carefully to serve as an edge guide, and the piece to be cut should be considerably longer than the desired product. The workpiece is slowly advanced through the blade along the guide, but in a direction perpendicular to the direction of the blade's rotation. Using stock longer than the needed piece (but it should have the correct width) will help the worker keep fingers and cutter away from each other.

Photo 5: The wooden mounting disks which help attach the tube to the yoke. Note the curved undersides and the angled hole for the machine screws. Note also the recesses for the machine screw heads, which permits the washer to present a flat surface to the inside of the yoke. Note the center hole is threaded to accept a 1/2" bolt. (A 1/2" tee nut would accomplish the same thing.) The pair of 1/2" bolts, which extend through opposite ends of the yoke at the top of the 'U' and thread into these washers, form the declination axis for the telescope.

Photos 6 and 7: The yoke, made from three thicknesses of 3/4" plywood, mounts to the tripod with the 1" dowel which serves as the polar axis. The dowel is driven tightly into a hole bored in the bottom of the 'U'. On opposite sides of the yoke are two 1/2" holes drilled to accommodate the bolts which form the declination axis.

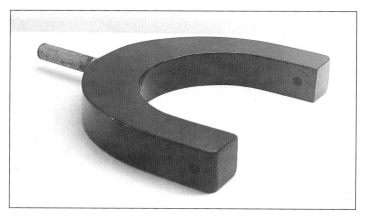

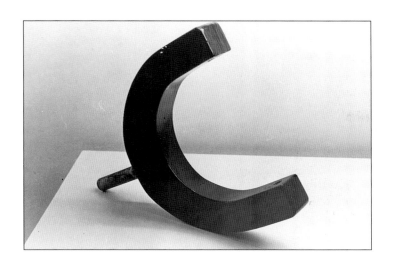

The outer width of the yoke in not critical, but the inner dimension is; this gap must allow a snug fit between for the tube with its wooden washers in place. Paraffin is a good lubricant for wooden surfaces which must move smoothly on one another, such as the outer surface of the tube's mounting disks, the polar axle, and the tripod legs.

Photo 8 : The finished telescope. The telescope is normally stored in two parts, the tube in its yoke and the tripod. Assembly is quick and easy: simply extend the telescope legs and insert the polar axis dowel into the tripod head. Aim the polar axis toward Polaris (if you live in the Northern hemisphere) and you're ready to observe.

An Eight Inch f/5 Reflector

Introduction

The eight inch telescope described in this project has a focal length long enough to make it useful at fairly high magnification, and thus ideal for lunar and planetary observing. But it's even better for viewing extended objects - galaxies, clusters, etc., for which fairly low power is preferable. Few telescope designs are ideal for both kinds of observing: one usually has a particular purpose in mind when choosing a telescope design, and hybrids are often unsuccessful. My own observing interests incline more toward objects such as galaxies, nebulae, and clusters, which are best observed with a telescope combining a wide angle of view, large light gathering power, and low magnification.

Bob Miller and I made an eight inch f/5 mirror some years ago, and I wanted to make a convenient and easily transported mounting for it. I decided to make it an equatorial mounting, and portability requirements dictated that the entire mounting be easily disassembled. Further, not having a machine shop available, I decided to design a mounting which used 2" pipe fittings for the polar and declination axes; motion on these axes was to be accomplished by rotating the axis on its pipe threads.

Many would object that pipe threads are unsuitable for this purpose because they are inherently either too wobbly, or they're too tight and free motion is impossible. This is all true — sort of. Clean pipe threads would certainly possess this disadvantage; the trick is to get them all clogged up with old paint and dirt, so the fittings will turn readily but fit snugly, and wobbling is discouraged. It takes a little practice to hit upon the right combination of paint and dirt, but I found that black spray paint does well, and its application is easily controlled. Dirt is easy to find.

Making the mounting

Designing a simple and easily transported tripod took some time. I wanted something that would readily disassemble and fit into the trunk of a car, along with the tube and other telescope apparatus. Above all, it must be sturdy. It finally occurred to me that I could combine these features in a design in which the tripod, shaped essentially like a pyramid with a triangular base, had a removable polar axis forming one leg of the pyramid and a simple triangular frame forming the other two. The pyramid would be formed by joining these two pieces. In other words, one leg of the tripod would be the polar axis, and the triangular frame would form the rest of the pyramid.

The polar axis was made from a piece of 4" x 4" pine, with a short piece of 2" x 4" pine extending perpendicular to the 4" x 4". This tongue is constructed to slip into a similarly shaped opening at the top end of the triangular tripod base. This protrusion is what holds the whole tripod together, so it must fit snugly into its opening (but not too tightly, or it'll never come apart again!) Obviously, considerable care must be taken cutting and fitting these pieces, in order to assure a smooth slip fit.

The entire telescope is assembled in the field without any fasteners. Each of the main components fit together and are held tightly by their own weight. Years of experience losing wing nuts and other fasteners in the grass have demonstrated the advantages of this design over those which bolt together. Photos 3 and 4 show the triangular base, with its opening for the polar axis and its 2" x 4" tongue. The triangle, built from 30" long pieces of 2" x 4" pine, is glued and bolted together for extra strength.

The desired length of the polar axis must be adjusted for the observer's latitude. In this example, in which the telescope will be used at latitude 40 degrees north, the total length of the 4" x 4" is 36", of which 24" extends below the tongue.

PARTS CHECKLIST FOR THE 8" f/5 REFLECTOR

A. Telescope tube assembly
1) Aluminum tube 10" ID x 45" long. Fiberglass tubes also work well.
1) Finder scope.
1) Eyepiece holder and assorted eyepieces. We use a rack and pinion eyepiece focuser on this telescope.
1) Elliptical diagonal mirror 1.32" and holder
1) Aluminized 8-inch mirror, approx. 40" focal length. Common commercially-made mirrors are available is 48" inch focal lengths. If you use one of these mirrors, be sure to use a tube 8 inches longer.
1) Primary mirror cell.
1) Carrying handle and mounting screws.
1) Pair of aluminum end rings to fit tube.

B. Tube saddle and saddle support:
1) Piece, 1" thick oak (or other hardwood), 8" x 10".
2) Pieces, 1" oak, 2" x 8" cut with curve to fit tube.
6) 1/4" x 20 interior threaded wood inserts to mount tube to saddle.
6) 1/4" x 20 x 3/4" long round head bolts.
2) Pieces self sticking felt pads to cushion tube. 6 2" long #10 wood screws to attach curved oak pieces to saddle.
Strong wood glue.

C. Polar axis:
1) Piece 4" x 4" pine, 36" long.
1) 1" Oak, 6"x6", my be cut into an octagonal shape (optional).
2" Pipe fittings: 1 tee, 2 flanges, 2 close nipples (2" long), counterweight attached to 2" pipe. [The counterweight used in this telescope is a 2" pipe, 20" long, with a cap at the end. The lower 14" is filled with lead.
1) Piece 2" x 4" pine 6" long.
3) 1/4" x 2 1/2" Lag screws.
3) 1/4" Flat washers.
7 or more) 5/16" Tee nuts (count the mounting holes in the flanges and get 1 more.
6 or more) 5/16" Machine screws 1 1/2" long.
1) Knob to turn 5/16" x 2 1/2" machine screw.
2) Pieces 1" oak, 10"x12" to construct saddle support.

D. Tripod triangle:
6) Lag screws, 1/4" x 2 1/2" long.
6) 1/4" Flat washers.
1) Piece 2" x 4" Pine, 30" long (base of triangle).
2) Pieces 2" x 4" Pine, 26" long (sides of triangle).
2) Pieces 2"x4" Pine, 14" long (top supports for tripod triangle).
Paint (flat black for interior of tube), exterior varnish.

Polar and Declination Axes

At the north end of the polar axis is a piece of oak fitted with 5/16" tee nuts (see Photos 5 and 6). The tee nuts accept the bolts which attach the polar and declination motion axes. First above the oak polar axis head comes a 2" pipe flange; above the flange is a 2" close nipple attaching a 2" tee. Rotation of the tee on this close nipple provides smooth polar axis motion. Below the tee is a pipe which contains a counterweight; this pipe happens to be filled with lead, but a coffee can filled with concrete would do just as well. Above the tee is another 2" close nipple, to which is attached another flange. Rotation on the threads of this second 2" close nipple provides motion about the declination axis. Finally, attached to this flange is a wooden saddle support made of oak, to which the telescope tube attaches.

Making the Tube and Accessories

All the pipe fittings except the counterweight remain connected to the 4"x4" polar axis. In practice, the entire telescope disassembles into four separate components: the counterweight (which screws into the lower end of the tee); the polar axle (to which is also attached the pipe fittings forming the polar and declination axes, as well as the saddle support built to receive the tube assembly); the wooden triangle which helps form the tripod; and finally the telescope tube itself. The telescope is put together by first fitting the polar axle tongue into its matching opening on the triangle; this is followed by screwing the counterweight pipe into the tee, and finally by sliding the telescope tube saddle into the matching saddle support and giving the knob a twist. That's all there is to it: assembly takes no more than about 30 seconds. Because this telescope's regular assembly and disassem-

bly are accomplished simply by fitting matching parts together; there is no need for extraneous wrenches, wing nuts, hammers, bolts, washers, or sump pumps; that's the advantage. The disadvantage is that great care must be taken to machine accurately those wooden parts which must fit together in a slip fit, or they may bind and make disassembly much more difficult.

The saddle support design requires some discussion. Photos 7 and 8 show that the upper part, the tube saddle, has tapered sides which fit directly into the matching tapered sides of the lower piece, the saddle support. In order that these parts mesh cleanly without binding, you should cut all the parts from one piece of material. An oak stop cut from the same material is glued at the bottom end of the saddle support, and a 5/16" machine screw, attached to the saddle support through a tee nut, is tightened by turning a knob to provide just enough pressure against the tube saddle to assure stability when the telescope is assembled.

The tube saddle and the saddle support are matching parts: one is attached to the tube, the other to the declination axis. The upper half of the tube saddle has end pieces cut in a curve to fit the tube. Threaded into the curved end pieces are inserts with 1/4" x 20 internal threads; these receive screws from within the tube, thus attaching the tube to the saddle neatly and inconspicuously. I gilded the lily by putting self-sticking felt on these curved surfaces to cushion the tube.

Many of the parts which form this telescope were purchased commercially. The aluminum tube (10 inches in diameter), the mirror cell, the diagonal support , the eyepiece holder and eyepieces, and the finder were purchased from University Optics in Ann Arbor, Michigan. The availability of high quality telescope components such as these, and at modest cost, are terrific advantages to amateur astronomers who lack access to a machine shop. Thus, purchasing the components and assembling a telescope of excellent optical and mechanical quality is within reach of most pocketbooks.

In most commercially purchased mirror cells, note the adjustment mechanism to align the mirror. The part which holds the mirror is spring loaded to allow accurate collimation adjustments, after which the mirror's position is locked in place by tightening the long screws.

Superficial dirt and dust on the mirror cause little harm; far more damage can be done by trying to rub it off. A small indelible mark placed at the exact center of the primary mirror helps to align the mirror exactly (and causes no deterioration of the image).

Accurate alignment of the elliptical diagonal is essential for a good image. To assist in this alignment, a small indelible mark may be placed at the location on the diagonal mirror where the telescope's optical axis must intercept the diagonal. This is not at the center of the diagonal! Information about this detail is presented in the text. Also note the adjustment screws in the mounting brackets of the finder which permit precise alignment.

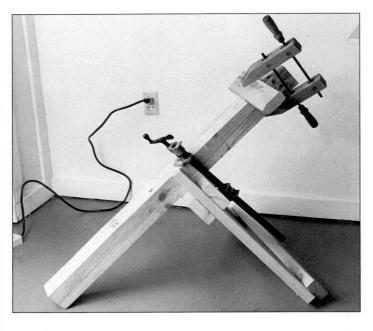

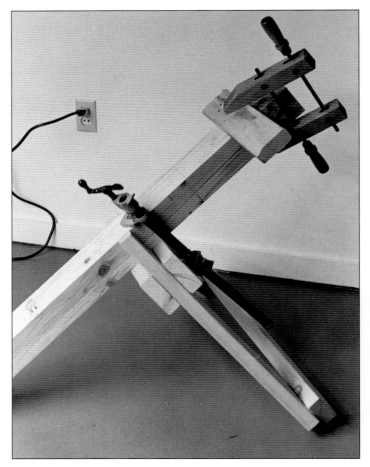

Photos 1 and 2: The parts of the tripod clamped together to indicate how to determine the relative lengths of the segments of the 4"x4" which forms the polar axis. The position of the polar axis 4"x4" relative to the tripod triangle must be adjusted so that the angle between the polar axis and the ground equals the observer's latitude. The short piece of 2"x4" scrap protrudes to the location of the telescope tube's eventual position when the telescope points vertically. Laying out these measurements correctly assures that the lower end of the tube won't strike the tripod when the observer is viewing overhead objects.

Photo 3: The tripod triangle, showing the opening which accepts the polar axis. The triangle is constructed from pine 2"x4", fastened together with 2 1/2 " lag screws and glue.

Photo 4: Exploded view of the polar axis before its assembly. Note that the tee-nuts and the lag screws have all been recessed so flat surfaces will face the parts which will later be attached.

Photos 5 and 6: The assembled polar axis showing the connecting 2"x4" tongue which fits snugly into the tripod's triangle, and the oak polar head to which a 2" pipe flange will be bolted. The tongue must be carefully machined to fit the opening exactly, or it may bind.

Photo 7: The oak saddle support, which attaches directly to the declination axis flange with the recessed tee nuts shown. Note the tapered sides, which will accept the matching tube saddle, and the stop block at the lower end. The tee nuts are recessed to provide smooth passage for the tube saddle.

Photo 8: The partly finished saddle support and the tube saddle, showing how they join, with the north end of the polar axle, the 2" pipe fittings which serve as polar and declination rotation axes, and the saddle support. Note the useless-looking adjustment knob: when tightened it thrusts a bolt against a flathead screw mounted on the underside of the tube saddle, helping to hold the telescope tube firmly in place.

Photo 9: Detail showing the threaded inserts mounted on the inner curved surfaces of the tube saddle's tube supports. These inserts accept 1/4" x 20 screws which are driven from inside the tube, thus mounting the tube to the tube saddle without conspicuous clamps or brackets.

Photo 10: The assembled mounting. The saddle support is permanently attached to the declination axis; it, plus the polar and declination fittings, and the 4"x4" polar axis, form one of the three parts which compose the mounting. The other two parts are the counterweight and the tripod triangle. The mounting easily disassembles, first by unscrewing the counterweight at the lower end of the 2" tee, and then by lifting the polar axle from the tripod triangle.

Photo 11: Detail of the mounting.

Photo 12:
Aligning the
finder scope
after the
telescope
has been
assembled
and
collimated.

Photo 13: Telescope on its mounting, pointing to the zenith. Note: The A-frame structure must be tall enough and fit on the 4 X 4 polar axis so the telescope will clear legs of the A.

A Ten Inch Dobsonian Telescope

Introduction

The smaller telescopes described in previous projects incorporate a variety of ideas, a range of construction materials and different an exciting step to a larger aperture telescope. This ten-inch telescope will make an exciting step to a larger telescope. It incorporates concepts proposed by John Dobson of the San Francisco Sidewalk Astronomers some twenty years ago. He visualized large aperture telescopes of simplicity and economy. These telescopes are not valued for their fancy mountings, motorized drives and sophisticated features, rather they are prized as "light buckets"; being affordable, relatively inexpensive, large aperture collectors of starlight. Currently, Dobsonian telescopes are commonplace at astronomy clubs and telescope-making conventions. This ten-inch telescope can be built during evenings over a span of a week or two at a cost of $500, including a good quality eyepiece.

The Dobsonian design is just a simple Newtonian telescope in the arrangement of the tube and optical components. The biggest difference is that rather than having an equatorial mount, such as those featured in projects four and five, where one motion of the telescope around the polar axis keeps the stars centered in the eyepiece, this telescope uses an alt-azimuth mounting. This works like a gun turret: the mounting rotates around a vertical axis (azimuth) and up and down (altitude). Such mountings are simple to make. However two motions are required to keep a star centered for prolonged viewing. Because these telescopes almost always have a short focal ratio (they work best at f/6 or less) and they work at low magnification, it generally takes little practice to nudge the telescope once a minute or so to follow the sky. The huge advantage is a low cost, easy-to-build, large telescope.

In recent years many companies noted for other kinds of telescopes such as Newtonians and Schmidt-Cassegrains have begun to manufacture telescopes of Dobsonian-style. Because of their simplicity, and lack of advanced features, they are quite inexpensive for their aperture. With such availability on the market, why build your own? These telescopes are easy to make, and require very few tools. Furthermore, with the vast array of excellent components available, you can built a telescope exactly to your needs and desires. There are many variations on this theme, in both design and building materials. We'll discuss possible variations (and features which shouldn't be changed) as we go.

Dobsonian telescopes of this type have enticed many more amateurs to build their own telescopes. Many amateur makers have built telescopes with apertures of 16 to 20 inches and even larger! More than twenty years ago, scopes of this size were virtually unheard of in amateur circles! In decades past, telescope makers frequently made their own mirrors, cast their own metal for mountings and rummaged through scrap yards for bearings and heavy metal parts. Fortunately, for most of us, this is no longer a necessity. Today some amateurs, still continue to make their own mirrors, but nowadays telescope makers have an excellent assortment of good-quality commercial optics available. Hard to make parts, like focusers and secondary mirror supports are readily available. Purchasing the few tough-to-make components and building the rest seems to be a good compromise to obtain a great custom telescope in a reasonable amount of time. Remember, making telescopes is fun and provides a great deal of satisfaction, however, we still would like to have time to use what we build!

Other ideas originally incorporated in Dobsonian telescopes were the use of light weight components such as plywood for mountings rather than steel, and rolled paper Sonotube™ tubes, rather than aluminum. Use of these materials was made possible for larger telescopes because of another innovation — light weight (thin) mirrors. Normally, Pyrex™ telescope mirrors have a thickness equal to 1/6 of their diameter. Some makers produce mirrors with a thickness of 1/10 the diameter or less. Such mirrors offer substantial weight saving for telescopes of more than 12 inches aperture. In this project we will use a "standard thickness mirror."

This project is for a ten-inch telescope with a mirror having a focal ratio of f/6, giving a focal length of 60 inches. The design for this telescope can easily accommodate a telescope tube with a shorter focal length mirror. If you make your own mirror, you can have exactly what you desire. Ten-inch commercially made mirrors are most commonly available with focal lengths of 45, 56 and 60 inches. This design can be scaled up fairly easily to 12 or 16 inches aperture or down to 6 or 8 inches. A telescope, like the one featured here, is easy enough to make, that it can be done successfully as a first telescope!

If you decide to use a commercially-made mirror, once you have determined your requirements and placed the order for the mirror, you don't have to wait until it arrives before constructing the telescope. Commercial optics may have a variation from the specified focal length of two or three percent, sometimes less. You need to know the focal length of the mirror, but a variation this small can be accommodated easily when the mirror is ready to be mounted in the finished telescope.

PARTS LIST FOR THE 10-INCH TELESCOPE

A. Telescope tube assembly:
    1) Sonotube™ 12" inside diameter x 72" long. Fiberglass tubes also work well. See Note 1, below.
    1) Low-profile eyepiece holder and at least one good quality eyepiece.
    1) Aluminized 10-inch mirror, approx. 60" focal length.
    1) Elliptical diagonal mirror, 2.14 inch minor axis and diagonal support.
    1) Primary mirror cell, if you decide to purchase one, or build the one featured in the text.

B. Mounting:
    1) 4' x 8' sheet of 3/4 inch A/B grade Fir plywood. The largest single piece needed is less than about 18 inches square, so for transporting, you may ask at the lumber yard to have this sheet cut into two 4' squares.
    1) Piece Formica™ 18" square. Visit a lumber yard which produces cabinets, or counter tops and ask for a sink cutout. Typically these are 3/4 inch thick plywood or particle board with Formica bonded. Such cutouts are sold for a couple dollars (or even discarded as scrap). Formica which is slightly textured works best.

1) Block of Teflon 4 x 6 inches, 1/2 inch thick, and enough to make 6 pieces: 1" x 1/2" x 1/8 inch thick. See Note 2, below.

2) Pieces of PVC tubing, 8-inches inside diameter, 1 inch wide. For this telescope 6-inch diameter tubing will work well. These will become the altitude bearings. See Note 3, below.

12) #6 Round head 1 1/2 inch long wood screws for attaching the altitude bearings to the tube support box.

3 dozen approximately) #10 x 1 1/2 inchFlat head wood screws.

2) dozen #10 x 1 1/4 inch Flat wood screws.

1) 1/2 inch or 5/8 inch Bolt, 3 1/2 inches long with two large washers and two hex nuts.

1) Bottle of wood glue.

C. Mirror mount:

1) Commercial mirror mount made of cast aluminum. In this project we'll make our own with the following parts. Much of the mirror mount will be made of plywood remaining from constructing the mounting.

6) Round head wood screws, 1 inch long with washers.

3) Flat head machine screws 1/4 inch x 20, 3 1/2 inches long, with hex nuts.

3) 2 inch Stiff springs which fit around the screws, above.

3) 1/4 x 20 Thread wing nuts.

3) Round head wood screws, 2 1/2 inches long with cap nuts.

6) #8 x 3/4 inch Flat head wood screws.

1) Piece of hardwood, 12" x 1 1/2" x 1/2" thick. A scrap piece of oak flooring is ideal.

1) Piece of hardwood, oak or hard maple, 12" x 1 1/2" x 3/4".

D: Secondary mirror support:

1) Commercial diagonal mirror support or "spider". Generally, this is regarded as the hardest part to build, so many amateurs choose to purchase this part. If you have access to a drill press and a hole cutter, making a secondary mirror support isn't really that difficult.

Additional parts for making your own secondary mirror support are:

1) Piece of moderately hard wood 7 x 10 inches, 3/4 inches thick.

Soft maple works very well here.

1) Piece of stiff aluminum, 2 inches longer than the tube diameter, 2 1/2 inches wide. This piece should be 1/16 inch thick, or greater, for adequate support.

1) 1/4 x 20 Bolt, 2 inches long with two washers and a hex nut.

3) #10-24 Round head machine screws, 1 1/2 inches long with cap nuts.

4) #6 x 1/2 inch Round head wood screws, for attaching the mirror holder to the aluminum support.

Note 1: Sonotube™ is usually available in stores which sell to builders and cement contractors.

Note 2: Teflon is occasionally sold in small pieces in hobby shops. Otherwise look in the phone book under "Plastics". Teflon is rather expensive, about $12 to $15 for such a piece, but fortunately there are several excellent alternatives. One of the best is a new ultra high molecular weight plastic (UHMW); it is much less expensive, very easy to cut and drill. It is tough and very slippery and make an ideal bearing surface. This piece will be cut into blocks, mounted to the ground board (see Figure 1) as the azimuth bearing surface on which the base board (Formica™) rides. Our telescope uses another similar material called Delrin. Teflon or UHMW is preferred for larger telescopes. Teflon is a trademark of Dupont, but here, we will use "teflon" to mean any of these materials.

Note 3: If you cannot find some scrap pieces of PVC tubing, don't purchase a 10 foot length for one telescope. Go to a hardware store and look for PVC as a "closet flange" or get a PVC pipe joint, cut an inch from each end. The PVC material for these bearings should cost just a few dollars. Other creative ideas have worked: bearings made from accurately cut wooden circles, surrounded on the outside by a strip of Formica™, or a metal strip (such as from a discarded tape measure, or ring cut from a coffee can, or even a 35mm film can).

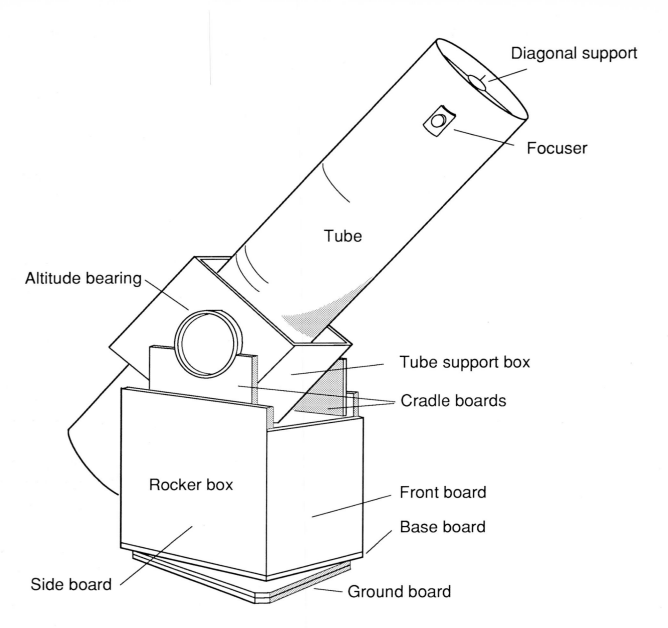

The Dobsonian telescope

Preparing the Tube and Tube Support Box

The cardboard Sonotube can be cut easily with a hand saw or on a table saw. Sonotubes are often scruffy looking and may have damaged ends, so select a tube carefully and allow for several inches to be cut from the ends. To cut on a table saw, have a friend support the far end, and slowly rotate the tube against the saw blade. The length of the tube should be at least 4 inches longer than the focal length of the mirror, so you need to know your mirror's focal length approximately. In our case we trimmed the tube to 64 inches.

Because Sonotubes are usually made as forms for concrete, most of them come with a waxed paper inner layer. Prior to painting, this paper must be removed. Just peel one edge loose and pull. The paper is glued in a long spi-

132

ral, so a steady pull will unwind it from the tube.

Choose a point on the tube for locating the eyepiece focuser. Ideally, you will want at least six inches of tube in front of the eyepiece to shield the eyepiece adequately from stray light. Using a keyhole saw, or a hole cutter in a drill, make a 1 1/2 inch hole. With the focuser as a template, center it on the hole then mark and drill holes for the mounting screws.

The actual diameter and length of the tube, the desired eyepiece height from the ground and the balance point on the tube all influence the dimensions of the mounting. We'll give the dimensions of the materials cut for this project and show how to alter them to suit a somewhat different telescope. You will likely want to make a sketch of the parts, showing their dimensions, if you choose to build a different instrument.

Referring to the drawing, if your telescope is dimensioned according to this projects, cut the following pieces:

Tube support box:

2 sides	12" x 12"
2 front and back	13 1/2 x 12:

Rocker box:

2 sides	15 1/4" x 18"
1 front	16 3/4 x 18"
2 cradle boards	12" x 14"
1 base	15 1/4" x 15 1/4"

(The base is the Formica or sink cutout, which will have the Formica side face down.

Ground board:

2 pieces	15 1/4" x 15 1/4"

On the cradle boards, cut a semicircle in the top, with the center of the circle coinciding with the top of the board (or even an inch above it).

The radius of this circle should be 1/4 inch larger than the outside of the PVC rings.

The tube support box is the first wooden part constructed; the other parts are made to accommodate it. Our Sonotube has a 12 inch inside diameter, and 1/8 inch wall thickness, giving a 12 1/4 outside diameter. The inside dimensions of the mirror support box should be 1/4 inch smaller than the outside diameter of the tube. If the support box is larger than this, the tube will deform and slip in its box. We want the tube to fit tightly and not require screws through the tube to keep it from slipping. For these dimensions, cut two sides of the mirror support box 12 inches square. Cut the other two sides 13 1/2 x 12. (These longer pieces will overlap the shorter sides.)

Drill and screw together gluing one longer side to the two shorter ones. Plywood is very soft, so for the screw threads to hold, the hole should be fairly small, about 3/16 inch. The piece being attached should be drilled large enough the the screw's shank to slip through. Countersink the screw heads. Use four #10 x 1 1/2 inch screws per joint. Put the tube in the support box and with the remaining long piece, drill and screwed it in place, without gluing, to permit removal of the tube.

Mount the two PVC altitude bearings onto the opposite sides of the tube support box, as shown in Photo 1. If you have trimmed two PVC rings, 1 inch wide, drill them in six equally spaced, 60 degrees apart. Since the PVC rings are relatively thin walled, exercise care to have the drill holes perpendicular to the face of the rings. An imaginary line through the centers of the mounted rings on opposite sides of the support box must be perpendicular to the long axis of the tube. Photo 2 shows the fourth side being attached to the mirror support box.

133

Photo 1

Photo 2

Constructing the Mounting Base

If you choose different dimensions, for example, a ten inch telescope with an f/4.5 mirror will have a tube which is 15 inches shorter. The height of the rocker box could be reduced by about 7 inches. For smaller telescopes or short

focal length instruments, you may prefer to increase the height of the rocker box in order to keep the eyepiece at a comfortable distance above the ground. For small departures from this design, the size of the side boards and cradle boards will change very little, if at all.

For telescopes of different aperture, 3/4 plywood will be good for 6 and 8 inch telescopes too. (Half-inch plywood would work for lightweight 6 inch telescopes. For 12 to 16 inch diameters, the side boards should consist 1" thick or 1 1/2" (two thickness of 3/4 glued together). The ground board should always be double thickness.

The critical measurement is the width of the front board and the size of the base and ground boards. These dimensions are determined by the separation between the cradle boards which should be 1/4 inch wider than the mirror support box, so the telescope will comfortably set down into the cradle without rubbing against it.

Photo 3

For example, our mirror support box is 13 1/2 inches wide, so a separation of 13 3/4 inches between the cradle boards is required.

Each cradle board is 3/4 inches thick so their outsides must be 2 times 3/4 + 13 3/4 or 15 1/4 inches. The side board are also 3/4 inches thick, so their outer faces will be separated by 2 times 3/4" + 15 1/4" or 16 3/4 inches. The size of the front board must then also be 16 3/4 inches wide by 18 inches high. An optional nice touch is to use a saber saw to cut hand hold holes on the three sides of the rocker box, as shown in Photo 3.

Photo 4

Once the final dimensions have been determined and the pieces cut, mount the cradle boards to the inside of each of the two side boards. The bottom of the cradle boards will be about 10 inches from the bottom of the

side boards. At this point, you have room for a few inches of vertical adjustment. Using #10 x 1 1/4 wood screws, about six per side, from the inside of the cradle board glue and screw the pieces together. Drill, glue and screw the front board, connecting the two sides. Countersink the screw heads slightly below the wood's surface. In Photo 4, one side has been glued and held together with clamps while the screws are tightened.

Take the base board with the Formica on the bottom surface, and mark the center. Drill a hole in the center to accept the half-inch bolt (which will attach the rocker box to the ground board). Next fit the base board to the rocker box's bottom. Photo 3 shows the assembled base with the cradle boards attached to the inside of the rocker box; note the Formica surface on the bottom.

The ground board is generally made of two thicknesses of plywood glued together. The ground board is the same size as the bottom of the rocker box. Make the two pieces, glue and screw them together. Although the two board were screwed together after gluing, the clamps shown in Photo 4, held together two edges which were slightly warped. Drill a hole centered in the finished ground board for the half inch bolt. Countersink the bolt head so the head and a washer are flush with the bottom side of the ground board. You may want to saw off an inch or two from the corners of the ground board to help prevent stubbing your toes on the protruding corners in the night. Some have built similar corners with round ground boards, but making pleasing looking circles of that size is additional work.

If you have three pieces of scrap hardwood, make and attach three feet equally spaced around the bottom of the ground board. We made feet from 2" x 3" x 3/4" thick wood. From the teflon block, cut three pieces about 2" x 3" and mount them in a circle (about 12 inches diameter) on the ground board's top surface. Mount them with screws whose heads are countersunk well below the surface of the block. Don't be tempted to use the Formica piece on the ground board and mount the teflon on the bottom of the rocker box! Experience shows using the Formica face down keeps a clean surface bearing against the teflon. Don't change this!

Photo 5

Bearings with teflon against Formica and against the PVC tubing are ideal for telescopes because they permit small movements of the telescope without sticking and binding. Furthermore they provide just enough friction so

the telescope will stay where you aim it. Don't try to improve this scheme by substituting ball bearings, otherwise rather than observing, you will spend the night trying to hold the telescope still. In the slightest breeze your fine telescope will become an expensive weathervane.

From the remaining teflon, cut six pieces 1" x 3/4 about 1/8 inch thick and tack these equally space inside the 'U' of the cradle, with small brads. Countersink the heads below the surface of the plastic.

Assemble the mounting by attaching the rocker box to the ground board using the bolt and washers. Lock the bolt in by tightening two hex nuts against each other. Set the tube and its support box into the cradle to check smooth movements and to insure parts fit and work together properly.

Building the Mirror Mount

If you choose to use a commercially-made mirror mount, mount the mirror in it. Be sure to use some adhesive felt, or some pieces of cork to both cushion the mirror against the cell and under the metal clips which press against the outer edge of the mirror's front surface.

It is rewarding to make as many parts of your telescope as possible, including the mirror mount. There are some simpler mirror cells, but the one featured here is attractive, extremely rigid and permits plenty of ventilation in the tube. The larger the mirror the more important it is to have adequate ventilation around the mirror as it cools in the night air.

The first step in constructing the mirror mount is to cut two circular disks from some of the remaining plywood, ten inches in diameter (or the same as the diameter of your particular mirror). One disk will have the mirror attached to it, the other will fasten to the telescope tube.

Mark the center of each disk. Select one disk to be the mirror support disk and label it.

On the mirror support disk, at 120 degree increments and a radius of 3 1/2 inches from the center mark and drill holes through the disk to accept 1/4 x 20 x 2 1/2 inch flat head screws. Make these holes small enough so the screws will thread into the wood. Countersink these holes so the screw heads will be slightly below the surface of the wood.

Make three hardwood, "L" shaped brackets from the pieces hardwood. One side of the "L" which attaches to the bottom side of the mirror support disk should be three inches long; the other side of the mounting bracket should extend to the height (thickness) of the mirror. These brackets are illustrated in Photo 7, which shows the completed mirror mount.

Photo 6

Using the first disk as a template, mark through the three holes drilled onto the second (bottom) disk. Drill these holes large enough for the 1/4 x 20 screws to pass through easily. From each of these holes, drill three more holes 1 1/4 inches in to accommodate the round head locking screws. These must thread tightly into the

wood. (If this doesn't happen as planned, you can resort to using tee nuts pressed into the upper surface of this disk.)

Photo 7

From the 3/4 inch thick piece of hardwood, cut three pieces 1 3/4" x 3 3/4". These dimensions are not critical. Round one end of each block to fit the curvature of the telescope tube. Before attaching any parts to the plywood disks, you may want to use a saber saw and cut a 3 1/2 inch hole in the center of each disk. This step is optional, but it will help let air circulate. Spaced 30 degrees from the drilled holes, mount these blocks to the bottom side of the plywood of the lower disk. The curved ends of the blocks should protrude about an inch beyond the disk and closely match the diameter of the tube. Use two screws to attach each one to the plywood.

When you attach the completed mirror mount to the telescope tube, drill mounting screw holes through the tube, into the ends of these blocks. Hardwood is preferred for durability, as periodically, the mirror will have to be removed from the telescope for cleaning.

Attach the three mirror mounting brackets. Use two screws each into the bottom of the plywood disk. When drilling the holes for these screws, you may want to elongate the holes slightly, to permit small adjustments to aid in securing the mirror in the cell.

Photo 8

From the bottom side of the lower disk, insert the three locking screws, screwing them through the wood far enough the cap nuts can be attached. On the mirror support disk, attach the three mirror brackets to the bottom side. From the top side of the disk insert the flat head screws by screwing them all the way into the wood. On the back side, on each of these screws place a washer and hex nut, tightened against the wood. For aesthetics, the washers and hex nut could be countersunk using a half-inch drill bit. Place springs on each of these screws.

Now the mirror support disk can be mated to the bottom disk, by aligning the bottom disk's holes with the flathead screws (with springs). Pushing these disks together against the spring tension, place washers and wing nuts and tighten so the wing nuts are on securely. Your completed mirror cell will look like the one shown in Photo 10. (In this photo, the mirror is shown, but it is not yet finished). Wing nuts permit small adjustments for collimating the mirror in the

telescope by allowing moving it along the optical axis and tilting it so the mirror's optical axis coincides with the long axis of the tube. Photo 7 shows the completed mirror mount the rear.

The mirror cell may now be disassembled for priming and painting. Rather than hide the details of our mirror mount with paint, we chose to apply a few coats of acrylic finish. This stuff dries in about fifteen minutes, so several coats may be applied in one evening.

When you are ready to mount the mirror in the cell, use some pieces of adhesive felt to cushion the mirror. Drill small holes into the tops of each of the three mirror support brackets and use small (#6 or #8) wood screws and washers to retain the mirror in its cell. Be sure to use felt on the metal (or fiber) washers to avoid scratching the mirror's delicate surface. Photo 11 shows the completed telescope tube and mounting with the mirror mount installed.

The Secondary Diagonal Mirror Support

The support for the diagonal mirror is usually regarded as being the most difficult part to build, therefore most amateurs purchase a four-vane "spider," which is make of aluminum. In a typical one, each of the ends of the four vanes attach with screws to the telescope tube. The center has a holder for the diagonal mirror which permits small adjustments along the optical axis, and rotational and tilt motions for final alignment. If you purchase on of these, fit the metal collar around the secondary mirror and attach it to the spider. This assembly is ready to be fit to the tube.

If you have access to a small drill press, and perhaps a circle cutter, building the diagonal mirror support, as we've done here, is interesting and satisfying. There are several designs which work well for these mirror supports. Ours is a bit more complex than some, but it permits all the fine adjustments of the commercially-made spiders.

To construct the secondary support, first use a circle cutter and cut in from a piece of 3/4 inch wood such as soft maple, three circular disks equal in size to the minor axis of the elliptical diagonal mirror. (Ours is 2.14 inches.) Most circle cutters leave a 1/4 inch hole in the center of the disk, which we will use. Glue these pieces together to make a cylinder 2 1/4 inches long. After the glue has dried, use a 1/2 inch Forstner bit or spade bit and enlarge the central hole (left over from the circle cutter), but do not drill all the way through, leave about 1/4 inch thickness remaining (which has a 1/4 inch hole).

Using a miter box, or if you are trying this on a table saw, being very careful, cut a 45 degree face on this cylinder. The part we will use has the 1/4 inch hole on the end.

From a piece of 3/4 inch hardwood, cut a block 2 1/4" x 1 3/4". This will be glued onto another piece (preferably) of 1/2 inch thick wood, cut to 2 1/2" x 2 1/4". In the first block, mark the edge for gluing. In its center, drill a 1/4 inch hole. At 120 degrees increments around this hole drill holes to accommodate the #8 x 1 1/2 inch screws. Note: these holes must be small enough the screws will thread into the wood. Two of the holes will should be about 3/4 inch from the 1/4 inch hole, the third one will be near the block when it is glued on, so make it about 1/2 inch from the large hole. Before gluing these blocks together, parallel to the long dimension in the center of the block, make a few saw cuts, making a groove, about 1/4 inch deep, just enough to accommodate the head of a #8 machine screw. Glue to two blocks together to form an 'L' shaped bracket.

Prepare the aluminum support vane by cut-ting its length to 2 inches longer than the inside diameter of the tube, 2 1/2 inches wide. Mount the 'L' shaped bracket onto center of the aluminum support using the four #6 wood screws. Now, the secondary support may be assembled, by pushing the 1/4" x 20 bolt through the cylinder and through the matching hole on the 'L' shaped bracket. Insert the three machine screws and put on cap nuts. These will press against the top of the cylinder. By loosening the nut on the 1/4 inch bolt, the cylinder will move along the optical axis. The completed secondary support is illustrated in Photos 9 and 10.

Photo 9

Mount the diagonal mirror onto the 45 degree face of the cylinder with flexible silicone cement. For larger diagonals such as this one, it is a good idea to find a sheet of cork or rubber or linoleum on which to glue the back side of the mirror, the glue this to the wood. For a telescope to give an image which is free from astigmatism,

Photo 10

Photo 11

the secondary mirror surface must remain optically flat. Allowing a cushion between the glued surfaces helps the mirror survive temperature changes without undo strain.

Now fit this diagonal support to the tube. Remember that in a relatively short focal ratio telescope, the diagonal mirror is offset slightly from the optical axis, away from the eyepiece. (See Appendix A, for details.) In our particular telescope, this offset is a little less than 1/10 inch. This can be accommodated easily of offsetting the diagonal support in the tube. One half inch from the ends of the aluminum bracket, drill mounting holes in to the aluminum then, bend one inch tabs on the end and fit to the tube. Drill matching holes in the tube and mount each end with a pair of machine screws and nuts.

Final Assembly

Any parts which have been assembled should be disassembled for sanding and painting. Use a primer coat on the plywood and Sonotube. A metal primer is required to get paint to stick to the aluminum on the diagonal support. The diagonal support should be painted with a flat black paint. Latex is very dull and it cleans up easily. The inside of the telescope tube should be painted flat black.

Reassemble the telescope tube, its support box and mounting. Mount the secondary mirror assembly, remembering the mirror offset (away from the eyepiece) so the mirror appears approximately centered in the eyepiece focuser drawtube. Install the primary mirror in its cell. Make sure no heavy pressure is applied anywhere to this mirror, it should rest firmly, but comfortably without being squeezed. With our custom made mirror mount, use a small screw and felt covered washer on the ends of each of the three mounting brackets to retain the mirror against the cell.

Now we'll mount the primary mirror in the tube. We must be careful to get the spacing from the secondary mirror and eyepiece correct. We'll do this in two steps. Note the focal length of the mirror will be the distance along the optical axis from the center of the mirror's front surface to the diagonal to a point 1/2 inch above the eyepiece focuser's drawtube, when the focuser is adjusted all the way in. Calculate that distance. Allowing for the thickness of the mirror and and mirror mount, mark the approximate position of the mirror cell's mounting holes on the outside of the tube.

The recommended way to insure these mirrors are in correct relative position without first drilling holes in the tube (unless you have plenty of confidence), is to get a friend and take the telescope outside. Insert an eyepiece, and set the focuser's position so the eyepiece is midway in its range of travel. Aim the telescope at some distant object, at least several blocks away. Have your assistant slide the mirror in its mount so the mirror cell's mounting brackets are approximately lined up with the calculated marks. Have the assistant move the mirror cell along the tube in small increments so you will find an image which is nearly in focus. Don't expect perfect images, yet, since the optics are not collimated. One the proper point is found for the mirror cell, mark this location and drill the mounting holes through the tube and into the mounting brackets. (With a commercial mirror cell, just drill through the tube!) Now, the mirror can be installed, and the telescope is ready to return outdoors for final collimation, as described in Project 4.

With a telescope of this size, adding a finder scope is a good idea. Many relatively inexpensive finders are available, typically with apertures of 50mm or so. These are often mounted with mounting rings screwed onto the telescope tube. You may even find some lenses from a surplus store and build your own small refracting telescope as a finder. We have seen some made from lenses from photo copiers.

At a cost of less than $1, you can find a pair of eye screws and mount them on the ends of a narrow board, about 10 inches long. Mount this onto either the telescope tube or tube support box, and you will have a effective gun-sight finder. Be sure to mount this with a provision for some adjustment so the finder can be aligned with the telescope.

This telescope should be rugged and nearly maintenance free for years, providing a great deal of observing pleasure!

A Scope Like "Alice"
Notes on the Construction of a Portable 8" Dobsonian Telescope

Setting the Stage

In 1991, Bob Scholtz (my brother-in-law) and I took a trip to Australia to spend some time observing with the members of the Astronomical Society of New South Wales, and to trek through Australia's Northern Territory... the fabled "Outback."

Bob and I are both amateur telescope makers (ATM's) from way back, and both of us have multiple instruments under our belts. So it was only natural that we decided to build a special telescope for the trek Down Under. Our plan was to put together a telescope that could be transported to Australia without taking up our entire luggage allowance. I had recently come across glass blanks for making a 6" telescope mirror, and planned to ask Bob to grind and polish the optics for a portable telescope I would build. I knew that Bill Herbert, another Columbus ATM, had a nice focuser he wasn't using, so I asked if I could borrow it for the trip. Bill agreed, and the next Saturday morning I found myself in his basement collecting the focuser and asking him why he had it in the first place.

Apparently he had planned to build a rich-field 8" telescope but never finished it. However, he still had the completed mirror there in his basement. When Bill learned I was planning to build an ultra-portable telescope to take to Australia, he gave me the 8" f/4.3 mirror and the 2" focuser. (So much for having Bob make a 6" mirror!) Bill then spent an entire month refiguring the mirror and then had it coated with enhanced aluminum as his contribution to our trip.

I set about making the rest of the instrument. My goal was to design and build a telescope whose optical assembly could be transported as airline carry-on baggage. I selected a two truss tube design with an lengthy pedigree; the new telescope drew ideas from telescopic creations of other ATM's, including Thane Bopp, Tom Burns, Dick Suiter and Bob Bunge.

Dave Kriege of Obsession Telescopes contributed suggestions for finish and materials. I used stained Baltic birch overcoated with polyurethane for the structure, PVC pipe for the altitude bearings, Wilsonart's "Ebony Star" laminate for the azimuth bearing, and etched virgin Teflon for bearing pads. Decorative wood inlay strips on the corners were added to cover the screws that were used during construction.

The finished mirror box contains all the optics and focuser, and stores in the overhead compartment on most airlines. (It also "fits" under the seat in front if you don't mind having some of it stick out and keeping your feet around it so the flight attendant doesn't notice.) When stored, the truss tubes attach to the inside of the tripod legs. Thus, there are only the two pieces to the whole instrument. The tripod was designed to travel in a duffel bag along with sleeping bags and other soft materials, and can be checked as normal baggage on the plane.

As a finishing touch, I added to the mirror box cover a raised-relief map of Australia cut from contrasting wood. The telescope was then named Alice for Alice Springs, one of the Outback destinations on our planned journey.

Alice was started in November 1990 and finished the following April, less than a week before we headed south.

Now for the Details

Dobsonian telescopes work. My plan with Alice was to make a telescope that would be easy to use and interesting in appearance, but that would not vary significantly from the basic principles of Dobsonian design.

Some of the specific problems encountered and solutions developed in that process are described below. I can claim little originality in my telescope, as most design solutions are hardly more then variations of ideas developed by other amateurs.

I designed *Alice* with the following parameters in mind:

- **High Quality Optics** - Even though the telescope was technically a Rich Field Telescope (RFT) and supposedly good for lower magnification views only, I wanted *Alice* to be as high optical quality as possible.

- **Optimal Performance** - I decided to do everything possible to make the telescope's performance everything it could be. That meant evaluating the entire system, including optical supports, the diagonal, the focuser and anything else that came to mind.

- **Stability** - Despite its portability, I wanted my telescope to perform like a classical Dobsonian, providing the smooth, backlash-free motion for which the design is known.

- **Portability** - *Alice* not only had to be an excellent performer, but also had to be extremely portable. The mirror box and focuser needed to fit in the overhead compartments on commercial airliners, and the rocker, truss tubes and tripod needed to fit in a duffel bag that could be checked as baggage.

- **Ease of Use** - The telescope had to be easy to set up and use. Because assembly would be required whenever I moved the telescope, I wanted a design that would go together quickly, would not need extensive collimation in the dark and would not have any loose knobs, bolts or wing nuts to fumble with cold fingers and drop onto the ground (or the mirror!).

- **Appearance** - As Louis Sullivan said a century ago, "Form ... follows function." In other words, make a telescope that works first, and then make it pretty. Thus, once the above five criteria were addressed I would be free to design a telescope that would also look good. *Alice* was to be the first telescope I had built that would

not be painted. No hiding behind wood filler and an extra coat of paint this time!

The Track & Truss Solution

After mulling over numerous telescope designs that would meet my criteria, I decided to go with a two-truss tube Dobsonian design based upon telescopes designed by Missouri ATM Thane Bopp and further developed by Dick Suiter, Tom Burns and Bob Bunge in Ohio.

In short, the telescope would be little more than a box-like enclosure for the mirror attached to a flat focuser board with two short truss tubes. The tubes would connect to both the mirror box and the focuser board by being drawn down into short channels by non-removable bolts. Thus, the alignment of the focuser board and diagonal would be fixed above the primary and assembly would produce nearly perfect alignment every time.

Because the telescope would be quite short (under 35" focal length), I decided to build a short collapsible tripod to support the Dobsonian rocker at a height that would place the eyepiece at a comfortable sitting height when the telescope was pointed toward the zenith. Likewise, when the telescope was pointed toward the horizon, the eyepiece could still be reached from the same seated position, although I would have to lean over a bit.

A little further figuring refined the design so that the focuser board and the diagonal assembly would store inside the mirror box, making the entire optical assembly a single portable package. The initial design seemed complete. I was ready to begin work.

Construction Materials & Tools

Alice is built of Baltic birch plywood, chosen because of its appearance, number of ply and lack of voids. It was certainly not chosen because of its price ... that stuff is like gold! I understand that since the breakup of the Soviet Union good Baltic birch has become harder to get. Apparently the "in" material is now Finnish birch. I'm told that the Finns sold their old mills to the Soviets and now produce their birch plywood with new equipment.

I added inlay strips to key places on *Alice*. The inlay accents not only look nice, but they also cover the many screws and other fasteners that help hold *Alice* together. Contrary to what most people expect, the inlay is easy to apply. Simply rout a shallow groove, put some wood glue on the prepared inlay strip, and set it in place. Later you can sand, stain and overcoat the inlay along with the rest of the wood. Inlay strips in a variety of patterns (and the wood handles and knobs used elsewhere on the telescope) are available from well-equipped woodworking outlets.

I didn't compromise on the strength or permanence of the telescope; all connections were glued and screwed. All plywood was stained with an oil-based golden oak stain and covered with several coats of gloss polyurethane.

Every time I build a telescope I learn how to use a new tool. In the case of *Alice* the tool was a router with a router table. Every circle, arc, groove and channel was done with the router. I had to use carbide tipped tools, as the Baltic birch plywood is full of glue and cuts very hot.

Besides the router, construction required tools no more sophisticated than an electric drill, a belt sander, a fine sander and a paint brush. Such simplicity of construction was important because I am not an accomplished builder, and I don't have a well-equipped shop.

Others helped me by providing me with access to a table saw (for starting out with square corners), a drill press (for installing threaded inserts), a band saw (for cutting the tripod pieces), and a scroll saw (for that map of Australia).

Mirror Box

The "heart" of the telescope is the mirror box; it contains (and stores) all of the optical components, and must be built with enough care to position the primary and secondary mirrors precisely and keep them aligned.

• **Box** - The box is made of 1/2" (seven-ply) Baltic birch plywood. Its outside dimensions are 10" wide by 9.5" long by 11.5" tall. The choice of the 9.5" by 10" cross-section was not arbitrary. That size allowed a 1/4" clearance on three sides of the mirror and a 3/4" clearance on the side toward the focuser. The larger clearance on the focuser side was necessary for the optical path to clear the blocks that support the secondary vanes (described below).

The choice of dimensions also allowed the focuser

board to be the same dimension as the side of the mirror box (9.5"), which in turn made it possible to center the truss tubes in the box corners. The centering was necessary for two reasons: (1) to allow the plane of the focuser board to be the same as that of the mirror box side, thus allowing the focuser to be as close to the telescope's center line as possible, and (2) to allow the channels for the focuser board attach to the truss tubes from the side and the mirror box channels to attach from the top and bottom.

• **Cover** - I also built a plywood cover to fit over the front of the mirror box. The cover is held in place by three cupboard magnets and pops off easily when the cover is pulled by its two attached handles. The cover attaches to three additional magnets on the back of the mirror box when the telescope is in use, where it both protects the mirror adjustment knobs and helps balance the telescope.

I also added a contrasting raised-relief wood map of Australia to the mirror cover. No real reason ... just thought it looked neat.

• **Centering Pads** - To keep the box centered in the rocker and to minimize any friction and abrasion when moving the telescope, small pads are attached on each side of the mirror box and inside the front of the rocker.

Mirror Cell

Cell Design - The primary mirror cell is modeled after the design used by Richard Berry in his excellent book, *Build Your Own Telescope*. Basically, the mirror sits atop a square of 1/2" Baltic birch plywood and rests on three 1" diameter pads of silicon rubber. At each corner of the square cell a short section of 3/4" dowel rises to the height of the front surface of the mirror. The dowels are split in half with the flat side adjacent to the mirror. The mirror is separated form the dowels by pads of silicone rubber squeezed through holes drilled in each dowel. The result is a mirror cell that holds the mirror securely, without any strain or diffraction-creating clips over the front surface.

Cell Adjustment - The cell rests on compression springs that surround three attached bolts that extend out through the back of the mirror box. Adjustment is done with phenolic knobs at the back of the telescope.

Mirror Cover - I made a Plexiglas cover that attaches to the primary mirror cell and protects the mirror from dust, dropped items and prying fingers. The clear cover has four Velcro spots that secure the Kydex light baffle during storage. Although it is removed for observing, the cover is usually in place when people see Alice in daylight. Many times I have been asked if the four black spots on the primary are some new type of optical "thing."

Transporting the Mirror - The cell is left in the box at all times; the mirror is not removed when the telescope is transported. The mirror box assembly may be heavier that way, but the mirror is as safe as could be and never gets handled during normal use.

Focuser Board

General Design - The focuser board supports both the secondary mirror and focuser, and stores in grooves cut into the inside surfaces of the mirror box. Basically, focuser board itself is very simple. It measures 9.5" wide by 7" tall and is cut from 1/2" plywood. One side supports the focuser and the other holds truss channels (described in the next section) and the secondary mirror assembly.

I cut a small hole in the top of the focuser board that acts as a handle for pulling the focuser board out of its storage position inside the mirror box. As a final touch, I mounted a small phenolic knob on the focuser board to serve as a handle for moving the telescope.

Support Vanes - The tapered secondary support vane was cut from 1/16" aluminum with a saber saw. I then filed the burrs off the edges and bent the piece to shape in a vice, clamped between two pieces of wood. (The wood keeps the aluminum rigid except at the point where you want to bend it.) A small hardwood block glued and screwed to the vanes supports the Novak secondary holder. Adjustment is achieved with thumb screws.

The support vane is bent so that the vanes intersect at a 90 degree angle, thus producing a traditional star image with four diffraction spikes at right angles. To achieve the correct angle in the space available, I had to mount the spider vane assembly on wood blocks attached to the focuser board. The blocks also act as stops for the truss tubes and allow me to place the focuser board in the correct position for tightening the attachment bolts.

Light Baffling - *Alice* has two features that baffle light and allow her to perform with a simple open truss: (1) a reducing plug inside the focuser; and (2) a light shield attached to the secondary holder.

• **Focuser Baffle** - All of the eyepieces I use are 1.25" diameter, including the 9mm Nagler that gets used almost exclusively (although it has a 2" barrel). I placed an aperture reduction plug in the end of the focuser tube closest to the diagonal, thus limiting the effective diameter of the 2" Tectron focuser to 1.25". There is no vignetting (except when I borrow someone's 2" eyepiece with a large field lens), and the area around the diagonal that can be seen through the focuser is drastically reduced, thus minimizing the size of the more important light shield.

• **Light Shield** - I cut a 6.5" diameter circle out of thin, black Kydex plastic and placed two strips of Velcro on the matte finish side. The strips attach to matching strips on the back side of the secondary holder. When Alice is assembled, the Kydex shield is removed from its storage position on the mirror cover (where it attaches to small Velcro dots) and stuck on the back of the secondary support. When the shield is in place, you cannot see past it through the focuser; the view is the same as if you had a complete telescope tube.

Is it as dark as it would be with a complete tube? No. But it's pretty good. I use the telescope in all conditions, and unless a direct light is shining past the shield and onto the inside end of the focuser, you would never know that the Alice is tubeless.

I am frequently asked if the light baffle causes much diffraction in the image, given that it is located directly in the light path. In fact, it does introduce an additional diffraction spike as a 45 degree angle to the four-vane cross caused by the secondary support vanes. In addition, the resolution on planets and double stars with the baffle in place is slightly less than with it removed, but I have never found it to be objectionable.

I have considered adding a "cage" that would create a light baffle outside the light path, but it always seemed like too much trouble. The current system, with its Kydex plastic disk attached with Velcro, is so simple that it would be hard to improve. Therefore, I have never pursued the addition of the extra assembly that would

be required to baffle without the extra diffraction. Even if a method were found, it wouldn't do anything about the diffraction caused by the secondary support vanes, the primary diffraction cause.

Twin Truss Tubes and Channels

Design Parameters - If *Alice* were to be portable, she couldn't have a traditional telescope tube; she would have to be a truss design. In keeping with my goal of making the telescope as easy to use as possible, I had several other objectives in mind:

• **No Loose Parts** - As much as possible, I wanted to avoid loose parts during assembly (they invariably get lost). That ruled out extra bolts, washers, wing nuts or special tools.

• **Minimal Recollimation** - I didn't want to have to do a major recollimation on the telescope each time I put it together. Once assembled, I wanted no more than a required "tweak" or two on the collimation knobs.

• **Quick and Easy Assembly** - Finally, I wanted a telescope that would be rigid before the focuser board was attached. I most certainly did not want a the truss tubes waving this way and that while I tried to secure a focuser board on top of them.

When I built my 17.5" telescope nearly a decade ago, I used eight tubes in a Serrurier truss. The tubes fit into split wood blocks where they were clamped with bolts. While the design worked extremely well (and has since been adopted by telescope manufacturers such as Obsession, Tectron, Jupiter, Starsplitter and AstroSystems), it is rather bulky for small telescopes. I wanted something more elegant for *Alice*.

Back to the physics books. Objects can move through space in six directions - up and down, back and forth, and from side to side. Each of these directions is called a "degree of freedom." To keep each truss tube in place, I needed reduce its degrees of freedom to zero. This could be achieved by pulling the truss tube into a groove and clamping it in place.

Because *Alice* was relatively small, I figured I could get by with just two parallel truss tubes. I placed them on the side of the mirror box for two reasons: (1) a side mounted focuser could be used while sitting down; and (2) placing both tubes in a vertical plane would increase the instruments rigidity by letting the tube work togeth-

er to offset the effects of gravity. (If you don't believe this, grab a piece of paper at one end and hold it vertically - it stays upright. Now take the same paper and hold it horizontally - it flops downward. The assembled truss acts the same as the sheet.)

Truss Tube Channels - I placed pairs of channels inside two corners of the mirror box and against the inside of the focuser board. The mirror box channels are attached to the top and bottom of the box (looking down into the telescope), and not the sides where the attaching knobs would hit the sides of the rocker.

To make the channels, I started with pieces of plywood 1/2" thick and 2" wide, and routed a groove 1" wide just over 1/4" deep down the center. The channels were then cut to 6" lengths for the mirror box and 4.75" lengths for the focuser board. I then added a small shelf at the lower end of the two 6" channels to provide a secure seat for the truss tube before tightening. Each channel was then drilled near its center to allow passage of the securing bolt.

Attaching the Truss Tubes - The truss tubes are held in place by carriage bolts that thread into inserts. The inserts are set into short pieces of dowel rod fitted inside the truss tubes. Tightening the bolts draws the truss tube down into the channels set in the mirror box and on the focuser board. The carriage bolts are permanently held in place by recessed lock nuts tightened so that they keep the bolts from wobbling, yet turn freely without binding.

To make assembly a "no-tool" operation, I attached knobs made out of wooden toy wheels to the carriage bolts and stained them to match the telescope. The knob diameters are 2" on the mirror box and 1.5" on the focuser board. Because there are no small parts, all assembly can be done while wearing gloves, an important point during Ohio winters.

I also placed two channel and knob assemblies on the inner surfaces of two of the tripod legs to secure the truss tubes when the telescope is disassembled.

Truss Tubes - The truss tubes are 1.25" diameter aluminum tubing with a .058" wall thickness. A single 72" section from a local hardware store provided just enough for both truss tubes and the center "slider" tube for the tripod.

Final Assembly - The most critical operation was

attaching the channels with wood glue; any inaccuracies would result in permanent "decollimation" of the telescope. My solution was to make sure all the pieces were cut square from the beginning and to simply snug the channels against the inside corners of the mirror box for approximate placement.

For final positioning, I glued the mirror box channels in place first (remembering to recess the channels slightly from the opening of the box to accommodate the mirror box cover that nests in the opening). While the glue was still wet, I loosely attached the truss tubes and adjusted them until they measured even and square. Then I tightened the lower assembly which clamped the lower channels in place. The next step was to glue the channels to the focuser board, tighten the upper assembly and wait for the glue to dry.

The Result - The telescope maintains collimation very well. I have disassembled the telescope, transported it several thousand miles and reassembled it without having to change a thing; the center spot on the mirror remains concentric and the finder crosshairs stay right on target. The most the telescope has ever required upon reassembly is a slight adjustment of the top two mirror cell adjusting bolts and a slight realignment of the finder.

Rocker

The side and bottom boards of the rocker that supports the telescope are made of 3/4" (13-ply) Baltic birch plywood. The front board is a single thickness of 3/8" plywood. All surfaces have circular holes cut out of them to reduce weight and provide handles.

The finished rocker is 11.75" wide and 10.25" deep, including the front board. The pivot point for the altitude bearings is 29" above the ground. The rocker's inside dimensions are just adequate to nest the mirror box for compact transportation.

Tripod

Traditional Dobsonian literature recommends building a ground board with three feet. Such a design would have made *Alice* a "knee-biter" telescope, and required the observer to crouch uncomfortably to use it. I'm too old for such contortions on cold (or warm) nights. I wanted a "sit-down" telescope, and that meant that *Alice* needed a tripod.

Stability was a must, and that dictated a short, stubby design with cross-bracing. When I built my first telescope over 30 years ago, I used a rigid tripod with a cross-braced center post. I own a very rigid photo tripod with a cross-braced center post. The excellent tripod design Richard Berry used for his refractor in his book *Build Your Own Telescope* used a cross-braced center post. Guess what I built for *Alice*?

Tripod Head & Slider Block - The tripod head is made of two layers of 3/4" plywood and has a circular center block that supports the slider tube (the piece left over after the truss tubes were cut). The slider block is also dual-thickness 3/4" plywood and is split to allow it to be clamped in place on the tube, thus locking the entire tripod assembly into a single, rigid assembly. A 1.5" wooden toy wheel serves as the clamping knob. The cross braces are made of 1/8" birch plywood strips obtained at a local hobby shop.

Tripod Legs - The tripod legs have an "I-beam" cross-section. The main plate of each leg is made of 1/2" plywood and the side pieces of 3/8" plywood. The legs are 19" tall and taper from a width of 4.5" to less than 1/2". Each leg is tipped with aluminum to prevent splitting and resist moisture absorption.

I routed a groove in each of the side pieces into which I inserted the main leg plate. The leg was then glued, screwed and clamped. The screw heads were then covered by an inlay strip. As mentioned above, I added channel and knob assemblies to the inner surfaces of two of the tripod legs to secure the truss tubes during storage. The third leg sports a wooden carrying handle.

Eyepiece Rack - As a final touch, I added a "Lazy-Susan" eyepiece rack to the slider tube. The rack has holes for seven 1.25" oculars and cutouts for two 2" accessories. The cutouts serve double duty, as they are necessary to fit around the attached truss tubes when the tripod is folded for storage.

Altitude Bearings

One of the main "secrets" of the Dobsonian design is the careful selection of bearing materials, the combinations I used being Teflon-PVC for altitude and Teflon-"Ebony Star" for azimuth. The 3/32" thick etched Teflon

I used came from scraps Dave Kriege sent me (although Ken Novak sells the same material), and the "Ebony Star" laminate was found at a flooring firm.

The side bearings are fairly large (6.5") to provide proper friction for comfortable observing. They are made of PVC gas transmission pipe donated by a friend. The side bearings are attached to the mirror box with screws on the inside of the mirror box.

I cheated a bit when it came to finding the balance point for the attaching the altitude bearings; I counterweighted the inside of the mirror box with some thin steel plates. So sue me! It looks better with the side bearings flush with the top of the mirror box.

Azimuth Bearing

Azimuth bearings on Dobsonian telescopes traditionally have been made of Formica or similar laminate sheets gliding on Teflon pads. Although references in telescope making literature have indicated that smooth, shiny Formica would not give the best performance, until recently no one had seriously tried to find the best materials available.

Then Wisconsin amateur telescope maker Peter Smitka tested various Teflon and laminate combinations for friction, and found that virgin Teflon and Wilsonart's laminate 4552-50 "Ebony Star" offered significantly less friction than other combinations. As a result, both the altitude and azimuth axes of his 200 pound telescope move with approximately two pounds of force.

Teflon comes in two varieties - virgin and reprocessed. Virgin Teflon is the first casting of the resin and contains no impurities. Reprocessed Teflon is made of remelted resins from previous uses and may contain impurities. Either will work for telescopes, but virgin Teflon will ensure the smoothest possible motion.

The azimuth bearing in *Alice* consists of three 2" by 1/2" by 3/32" thick virgin Teflon pads epoxied to the top of the ground board at 120 degree angles around the center. The pads are mounted 5" from the centering bolt, giving a bearing outside diameter of 10", roughly comparable to the diameter of the mirror (a good rule of thumb). By moving the pads in or out, friction can be increased or decreased slightly to "fine tune" the azimuth motion.

I attached a sheet of "Ebony Star" laminate to the bottom board with contact cement. A 1/2" centering bolt passes down through the bottom of the rocker and through the tripod head where it is secured with a nylon thread lock nut. Pointing the bolt down keeps the inside of the rocker clear, and allows minimum clearance and a shorter rocker.

No provision for changing the bearing friction is necessary in a telescope this size. The motion of the "Ebony Star" moving over the Teflon is smooth, vibration-free and without backlash.

Miscellaneous Points

Some of the other considerations in my telescope are set forth below:

• **Focuser** - Even though I would not be using 2" eyepieces (the 9mm Nagler is technically a 1.25" eyepiece with a 2" barrel.), I decided on a 2" focuser. I also wanted a short focuser to keep the telescope's diagonal size to a minimum. Because of the critical focusing demands of a short focus Newtonian (and because Bill Herbert gave it to me), I used Tectron's low profile focuser. The larger focuser also lets me insert a 48mm Lumicon Deep Sky or UHC filter into a 2" adapter and leave the filter in the focuser while I change eyepieces. (I also have a separate adapter without a filter, so I never have to screw and unscrew the filter in the dark.)

• **Secondary Mirror** - Size of the secondary mirror is a key factor in the performance of a Newtonian telescope. I decided to make mine as small as possible without adversely affecting performance. Unfortunately, in a telescope as "fast" as *Alice*, it just isn't possible to have a truly "small" secondary.

After poring over numerous articles on secondary size, quality, offset, obstruction and just about anything else, I sat down at a personal computer and developed a Lotus 1-2-3 spreadsheet to do the tedious calculations for me.

The model lets you enter the size of the primary mirror, its focal length, the outside diameter of the telescope tube, focuser height and desired fully illuminated field diameter. It then gives back the major and minor axis measurements of an "ideal" secondary. You can then enter the minor axis of the nearest standard size secondary (or the size you would like to evaluate) and the model will give you the actual fully illuminated field, offset from the optical axis, measured offset along the secondary's surface and the percentage of the primary's aperture and area obstructed. You can also view graphs of magnitude loss and overall field illumination as a function of distance from the optical axis.

I finally decided on a 2.0" minor axis 1/20 wave quartz secondary from E & W Optical. The secondary (actually 1.88" clear aperture in its Novak holder) creates a 0.26" fully-illuminated field and produces primary mirror obstructions of 25% (diameter) and 6% (area). The light loss at the edge of a one inch diameter field (the largest field stop on my Brandons) is 0.24 magnitude. With the 9mm Nagler vignetting is virtually nonexistent.

• **Finder** - The telescope is equipped with a Celestron 10x40 finder. The higher magnification makes it a better choice for locating objects in the "less than perfect" skies of my Ohio observing sites than the more frequently encountered 8x50 findersThe finder is mounted in twin alignment rings that rotate around the upper truss tube. The rotating assembly was necessary so I could move the finder out of the way when storing the truss tube in the rocker.

• **Collimation** - Proper collimation is critical in a fast focal ratio telescope. I use a set of Tectron collimating tools to check the optics each time I assemble Alice. In keeping with my "no-tools" philosophy, I replaced all collimating bolts and nuts with large knobs and thumbscrews. A slight turn of a mirror cell knob and a minor finder adjustment is all that's required to collimate the instrument.

• **Dew Prevention** - Dewing of optical surfaces is the enemy of any telescope with exposed optics - *especially* in the American Midwest. Because *Alice* has no tube to protect the optics, I sometimes encountered nights when the telescope was dewed up before she was completely assembled.

I solved the problem by installing the Kendrick Dew Remover system, developed by Canadian amateur astronomer Jim Kendrick. The system consists of heating elements that fit behind the secondary mirror, wrap around both ends of the finder, and fit around the eyepiece. The heaters are powered by a 12 amp-hour rechargeable battery, adjusted by a small electronic control box, and provide just enough heat to prevent dew from forming on the optics.

Now on bad nights the mirror box may be dripping wet, but the optics are completely dry.

• **Fan** - A telescope will not perform well unless it is near the ambient temperature. *Alice's* open structure facilitates cooling, but she can still take a half hour to completely cool. I accelerate the process by placing a small battery powered fan inside the mirror box and pointing it toward the covered primary mirror for a few minutes before each observing session.

• **Desert Storm Shield** - One of the slickest telescope accessories to come along in recent years is the Desert Storm Shield, an aluminized Mylar telescope cover manufactured by Columbus amateur Ralph Hoover. The aluminum coating reflects the sun's heat during the day, and the Mylar prevents moisture from reaching the telescope. Ralph graciously made a custom cover for *Alice* which travels with her everywhere.

One night at the South Pacific Star Party in Australia, an unexpected weather front brought a drenching rain that caught many telescopes on the observing field and lasted throughout the night. I awoke to the drumming rain on my tent many times that night, wondering how *Alice* was doing, covered only with a thin Mylar bag. No worries, mate. The next morning she was as dry as the Outback.

• **Traveling Cases** - *Alice* saw over 60,000 miles of travel in her first three and a half years, and she always traveled without extra protection - mirror box as hand luggage and the tripod in a large duffel bag stuffed with clothes. Although Alice was never damaged, by mid-1994 I figured that I had pushed my luck and tempted the luggage gorillas too often. I had custom shipping containers made that hold the telescope and accessories.

Now *Alice* travels first class; safe in her cases. Now if those gorillas can just keep from losing her ...

The Final Result

Well, how did everything turn out? I started out with the criteria of high quality optics, excellent performance, stability, portability, ease of use and appearance.

In fact, *Alice* is an excellent performer. When the telescope is temperature stabilized, images are sharp across a magnification range of 27x-349x. The 97x view through a 9mm Nagler is so good that I seldom use other eyepieces. The telescope is very stable and moves smoothly with fingertip pressure. The mirror box fits neatly in the overhead compartment of an airplane, and the tripod packs easily in a duffel bag, although I will use the shipping cases in the future. The telescope assembles in under two minutes and doesn't lose collimation from set-up to set-up. Finally, *Alice* is good-looking (at least to me).

I originally designed *Alice* as a special purpose telescope ... just to get some aperture on a plane. (I didn't design for maximum portability with resulting sacrifices in stability, as I knew from long experience that despite the murky skies in town I would use a telescope near home many times for each time I would transport it to a dark site.) However, *Alice* works so well that I use her almost exclusively. In the three years since her completion, we have accumulated over 60,000 miles of travel, including two treks through the Australian Outback and numerous trips to observing sites and star parties around the United States.

Between observing sessions, *Alice* stands unobtrusively in a corner of my den, waiting for the skies to clear. That's when her appearance helps. With Ohio weather, one spends more time looking at a telescope than *through* it.

Since finishing *Alice* in 1991, I have continued to pursue the twin-truss design, and recently completed six 10" f/5.6 instruments. The 10" version occupies a volume of two cubic feet plus the two truss tubes - overall, smaller than its 8" predecessor. Other ATM's have scaled the design up as far as 12.5" f/4.8 and it still works well.

A Final Note on Design

To the newcomer to telescope making, *Alice* may appear to be very different than most other Dobsonians. Such is not the case. For the most part, I restricted my design changes to cosmetic and non-critical areas of the telescope. A close examination will show that the main principles of the Dobsonian design such as matched bearing materials, large bearing diameters, short and rigid construction and no cantilevered components still hold for my telescope.

I have seen and used a number of Dobsonian telescopes in the past few years. Unfortunately, few of them demonstrated the stability that the design is capable of delivering. It seems that some telescope makers think that if they simply place a telescope on a plywood altazimuth mounting and throw in some Teflon and Formica at appropriate points, they have a Dobsonian telescope that will perform like the ones in the magazines. Not so. All of the things Russell W. Porter used to preach about overhang, small bearing diameters and flimsy construction are as true with Dobsonians as with any other style of telescope.

The Dobsonian approach is successful because it works. Materials, dimensions, construction techniques, loading and forces all play important parts in the design; the telescope builder who adheres to the basic "rules" of the Dobsonian can end up with a telescope that performs wonderfully.

Building Your Own

I am often asked two questions about *Alice*: Can I copy your design? Do you have plans?

My answers are: (1) yes, you can copy it; and (2) no, I don't have any plans (other than this article). However, it's not that simple. As I mentioned before, in designing Alice I drew on the ideas and techniques of a number of other amateur telescope makers. But I didn't "copy" anyone. Nor would I want someone else simply to "copy" *Alice*. The ideas are free for the sharing. After all, that's where I got most of my inspiration; other than the truss and track assembly almost every other design idea came from someone else.

Feel free to look over the telescope in whatever detail you choose. Use any idea that catches your fancy. But don't just copy what you see. Figure out why I did what I did and come up with a new solution that is better. Then take some pictures and write down what you did to help the next person. Above all, have fun!

"It is not ... unsportsmanlike to study closely the details of telescopes made by others and to 'lift' this or that feature from them, provided one improves these features."
— *Albert G. Ingalls — Amateur Telescope Making*

Correct Placement of the Diagonal Mirror

It was claimed in the text of two of the photo captions that the diagonal mirror's center should not be placed at the optical axis. At first glance, this may seem incorrect, but examination of Fig. 1 should be persuasive. The amount of offset is fairly easy to calculate. And as an added bonus, the calculation directly provides not only the value of the amount of the offset, but also the correct dimension of the diagonal mirror's minor axis. The values of these quantities are dependent on the mirror's diameter and focal length, the geometry of the telescope's optical design, and the distance of the focal plane from the optical axis. All this may sound unnecessarily complicated, but the concept is crucial for short focal length telescopes for which the amount of the diagonal's offset can be fairly substantial. Ignoring the offset will make the telescope impossible to collimate and seriously flawed images will discourage the observer.

That there must be an offset is obvious: the lower portion of the diagonal is closer to the primary mirror than the upper portion, and thus intercepts the converging cone of light rays at points where the cone is larger.

The exact equations describing the situation are rather messy, but the simple formulas given below provide results accurate to about 1%, certainly good enough for most purposes. The formulas also assume a field of view of 1 degree, a typical value. In these formulas, let:

r = mirror radius (half the diameter) in inches;

f = focal length of primary mirror in inches;

h = distance of desired location of focal plane from the optical axis;

m = ideal minor axis dimension of diagonal mirror (in inches);

d = amount of necessary offset in inches (see Fig 1).

These are the approximate formulas which allow one to find m and d:

$$m = 2rh/f + (f-h)/60$$
$$d = 0.7mr/f$$

Here are the values for several telescopes described in the text:

4 1/4" f/4	telescope: m = 1.13"	d = 0.09"
6" f/3.3	telescope: m = 1.37"	d = 0.14"
8" f/5	telescope: m = 1.32"	d = 0.09"
10" f/6	telescope: m = 2.20"	d = 0.12"

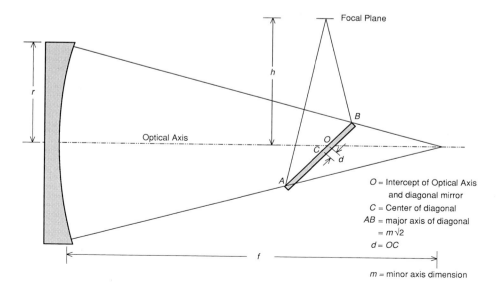

O = Intercept of Optical Axis and diagonal mirror

C = Center of diagonal

AB = major axis of diagonal
  = $m\sqrt{2}$

d = OC

m = minor axis dimension

FIG 1. The geometry of the diagonal mirror placement. Point C indicates the center of the diagonal, point O the intercept of the optical axis and the diagonal, and h is the distance between the optical axis and the focal plane. The major axis of the diagonal is indicated by the length AB, and d is the amount of required offset. Clearly, the line AO is longer than the line OB.

Basic Formulas

Magnification

The magnification, m of a telescope is given by:

$$m = \frac{f}{f_e},$$

where f is the focal length of the objective and f_e is the focal length of the eyepiece.

Focal lengths of telescopes are usually expressed in inches or centimeters, while eyepiece focal lengths are almost always expressed in millimeters. To convert between inches, centimeters and millimeters, 1 inch = 2.54 cm = 25.4 mm, and 1 mm = 0.03937008 inch.

For example a 10 inch $f/6$ telescope using a 16 mm eyepiece: 16 mm = 0.787 in. f = 60 in.

$$m = \frac{60}{0.787} \sim 76 \text{ x}$$

Exit Pupil Size

The exit pupil size, h is:

$$h = \frac{a}{m,}$$

where a is the aperture or objective diameter.

Minimum and Maximum Magnification

The exit pupil size must not be larger than a pupil size of a fully dark-adapted eye, which is about 7mm, otherwise not all the light coming through a telescope's optical system is being used. An exit pupil which is too large effectively reduces the telescope's aperture. As magnification decreases, the exit pupil size increases proportionally.

The formula for minimum magnification, m' is:

$$m' = 1.33\,a \quad \text{for } a \text{ in centimeters}$$
$$= 3.62\,a \quad \text{for } a \text{ in inches}$$

For example, using a 10 inch telescope, the minimum magnification is ~ 36.2, so the eyepiece focal length must be:

$$36.2 = \frac{60}{f_e} \text{ or } f_e = \frac{60}{36.2} = 1.657 \text{ in} = 42 \text{ mm.}$$

A 40 mm eyepiece is about the longest usable eyepiece for this telescope.

A good rule-of-thumb for the maximum magnification is about 50 x per inch of aperture. A number of factors usually limit the useful magnification per inch of aperture to less than this. If someone tries to sell you a 2.4 inch telescope that magnifies 500 times, run the other way!

Image Size

The size of the image of the moon at the focus: The moon subtends an angle of 1/2 degree or 30 minutes of arc, on the average. The size, h, of the moon's image is:

$$h = 0.0087\, f.$$

Resolving Power

All optical systems do not strictly obey the laws of geometrical optics. Even a perfect optical system will form the image of a point source (such as a star, because of the star's extreme distance) into a small image of measurable size, called the Airy disk. This phenomenon is called diffraction and it is caused by the wave-like nature of light. The angular limit of resolution from the Rayleigh criterion (after Lord Rayleigh) is:

$$\theta = \frac{1.22\,\lambda}{a} \text{ radians}$$

$$= \frac{1.22\,\lambda \times 206264.806}{a} \text{ arc seconds.}$$

where λ is the wavelength of light and a is the aperture. For yellow light, the color to which the eye is most sensitive, ($\lambda = 5.5 \times 10^{-5}$ cm), the resolution in arc seconds is:

$$\theta = \frac{1.22 \times 5.5 \times 10^{-5} \times 206264.806}{a}$$

$$= \frac{13.84}{a} \text{ for } a \text{ in cm}$$

$$\theta = \frac{5.44}{a} \text{ for } a \text{ in inches.}$$

A telescope a little larger than 5 inches is required to see features one arc second across on a distant planet. Under favorable conditions, a 12 inch telescope could resolve to better than 1/2 arc second. The book *Optics* by Hecht and Zajac, Addison Wesley, 1974 gives an excellent discussion on the topic of diffraction.

Limiting Magnitudes

The scale of magnitudes is based on the definition that a star of a given magnitude is the fifth root of 100 or 2.5119 times as bright as a star of the next higher magnitude. The larger the number, the fainter the star. A brightness difference of 100 times is 5 magnitudes. The faintest star we can see with the unaided eye is about magnitude 6.

The faintest star which can be seen in a telescope with an aperture a is approximately:

$$m \sim 6.8 + 5\log a \quad \text{for } a \text{ in cm, or}$$
$$\sim 8.8 + 5\log a \quad \text{for } a \text{ in inches.}$$

(Log is the common, or base ten, logarithm.)

Mirror Making Kits and Supplies:

Beacon Hill Telescopes
112 Mill Road
Cleethorpes
South Humberside DN35 8JD
England
0472-692959

Eftonscience Inc.
3350 Dufferin St.
Toronto, Ont. M6A 3A4
Canada
phone: 416-787-4581

Newport Industrial Glass, Inc.
1633 Monrovia Ave.
Costa Mesa, CA 92627
phone: 714-645-15000

Palomar Optical Supply
P.O. Box 1310
Wildomar, CA 92395
phone: 714-678-5393

Salem Distributing Co. Inc.
Box 25566 Manor Station
Winston-Salem, NC 27114
phone: 800-234-1982

Willmann-Bell, Inc.
P.O. Box 35025
Richmond, VA 23235
phone: 804-320-7016
catalog: $1 US

Mirror Coating Services:

Beacon Hill Telescopes
112 Mill Road
Cleethorpes
South Humberside DN35 8JD
England
0472-692959

Denton Vacuum
Cherry Hill Industrial Center
Cherry Hill, NJ 08003-4072
609-424-1012
609-424-0395 FAX

Evaporated Metal Films
701 Spencer Road
Ithaca, NY 14850
607-272-3320

P.A. Clausing, Inc.
8038 Monticello Ave.
Skokie, IL 60076
312-267-3399

Precision Applied Products
418 Rumsey Place
Placentia, CA 92670
714-738-4775

QSP Optical Technology
1712 J Newport Circle
Santa Ana, CA 92705
714-557-2299

Periodicals:

Astronomy
Kalmbach Publishing Co.
21027 Crossroads Circle
P.O. Box 1612
Waukesha, WI 53187
800-446-5489

Astronomy Now
Pole Star Publications, LTD.
P.O. Box 175
Tonbridge, Kent TN10 4ZY
(01732) 367542

Astro Shopper
P.O. Box 26
Ryde, Isle of Wight PO33 2XB
England

ATM Journal
The Amateur Telescope Makers
Association (ATMA)
17606 28th Ave. S.E.
Bothell, WA 98012

Sky and Telescope
Sky Publishing Corp.
P.O. Box 9111
Belmont, MA 02178-9111
800-252-0245

Starry Messenger
P.O. Box 6552-G
Ithaca, NY 14851
201-992-6865

Books:

*Advanced Telescope Making
Techniques,* Volumes 1 AND 2,
published by Willmann-Bell. ISBN
0-943396-11-5
& 0-943396.

*A Manual of Advanced Celestial
Photography*
Wallis & Provin,
Willmann-Bell, 1988

Astronomical Optics
Daniel J. Schroeder,
Academic Press, 1987.
ISBN 0-12-629805-X

Astrophotography
Barry Gordon,
Willmann-Bell, 1985
ISBN 0-943396-07-7

Build Your Own Telescope
Richard Berry,
Willmann-Bell.

How to Make a Telescope
Jean Texereau,
Willmann-Bell,
ISBN 0-943396-04-2.

*Microcomputer Control of
Telescopes*
Trueblood and Genet,
Willmann-Bell,
ISBN 0-943396-05-0

Prism and Lens Making
F. Twyman,
Adam Hilger, NY, 1988,
ISBN 0-85274-150-2

*Telescope Optics: Evaluation
and Design*
Rutten and van Venrooij,
Willmann-Bell,
ISBN 0-943396-18-2

Publication Sources:

Astronomical Society of the
Pacific
390 Ashton Ave.
San Francisco, CA 94112

Earth and Sky
256 Bacup Rd.
Todmorden, Lancs. OL14 7HJ
England
0706-817767

Kalmbach Publishing Co.
21027 Crossroads Circle
P.O. Box 986
Waukesha, WI 53187-0986
800-533-6644

Midland Counties Publications
Unit 3, Maizefield
Hinckley, Leices. LE10 1YF
England
(0455)-233747

Sky Publishing Corp.
P.O. Box 9111
Belmont, MA 02178-9111
800-253-0245

Willmann-Bell, Inc.
P.O. Box 35025
Richmond, VA 23235
800-825-STAR
Catalog: $1 US

Finished Optics:

A.E. Optics
28 Dry Drayton Industries
Scotland Road
Dry Dayton
Cambridge, CB3 8AT
England
0954-211458

A. Jaegers
6915 Merrick Rd.
Lynbrook, NY 11563, or
11 Roosevelt Ave.
Valley Stream, NY 11581
516-599-3167

Clear Star Optics
4193 Tallmadge Rd.
Roolstown, OH 44272
216-325-1722

Coulter Optical, Inc.
P.O. Box K
Idyllwild, CA 92349-1107
714-659-4621

D & G Optical
6490 Lemon St.
East Petersburg, PA 17520
717-560-1519

Dark Sky Telscopes
6 Pinewood Dr.
Ashley Heath
Market Drayton, Salop. TF9 4PA
England
063-087-2958

E & W Optical Inc.
2420 E. Hennepin Ave.
Minneapolis, MN 55413
612-331-1187

Edmund Scientific Corp.
N919 Edscorp Bldg
Barrington, NJ 08007
609-547-3488

Enterprise Optics
P.O. Box 413
Placentia, CA 92670
714-524-7520

Galaxy Optics
P.O. Box 2045
Buena Vista, CO 81211
719-395-8242

Iowa Scientific Optical
4231 Northwest Drive
Des Moines, IA 50310
515-255-0166

I.R. Poyser
15 Dale Rd.
Rochester, Kent ME1 2JP
England
0634 401220

Jon Owen
40 Birch Lane
Longsight, Manchester M13 0NN
England
061-257-3414

Lumicon
Suite 5
2111 Research Dr.
Livermore, CA 94550
510-447-9570

NOVA Optical Systems
P.O. Box 80062
Cornish, UT 84308
801-258-5699

Palomar Optical Supply
P.O. Box 1310
Wildomar, CA 92395
714-678-5393

Pegasus Optics
2301 W. orrine Drive
Phoenix, AZ 85029
602-943-3279

Precision Applied Products
418 Rumsey Place
Placentia, CA 92670
714-738-4775
800-575-4775

Star Instruments
P.O. Box 597
Flagstaff, AZ 86002
602-774-9177

United Lens Co.
259 Worchester St.
Southbridge, MA 01550
508-765-5421

Telescope Parts:

AstroSystems, Inc.
5348 Ocotillo Ct.
Johnstown, CO 80534
303-587-5838

Braithwaite Telescopes
9 Manse Brae
Dalserf, Lanarkshire ML9 3BN
England
0698-881004

Beacon Hill Telescopes
112 Mill Road
Cleethorpes
South Humberside DN35 8JD
England
0472-692959

D.I.Y. Telescope Store
89 Falcon Cresent, Clifton
Swinton, Manchester M27 8JP
England
0161-794-8650

Edmund Scientific
N919 Edscorp Bldg.
Barrington, NJ 08007
609-547-3488

Equatorial Platforms
11065 Peaceful Valley Rd.
Nevada City, CA 95959
916-265-3183
catalog $2 US

Khan Scope Centre
3243 Dufferin St.
Toronto, ONT M6A 2T2
416-783-7697

Kenneth Novak & Co.
Box 69
Ladysmith, WI 54848
715-532-5102

Lumicon
2111 Research Dr.
Livermore, CA 94550
415-447-9570

Motion Control Systems
P.O. Box 19632
Portland, OR 97280
503-244-0503
catalog $2 US

Orion Telescope Center
2450 17th Avenue,
P.O. Box 1158
Santa Cruz, CA 95061
800-447-1001 (ex. CA)
800-443-1001 (in CA)

Osborne Optics
139 Dean House
Eastfield Ave.
Newcastle upon Tyne, NE6 4UU
England
091-263-8826

Sherwoods
GT. Western Arcade
Birmingham B2 5HU
England
021-236-7211

Stardrive Systems
233 Bannock St.
Denver, CO
303-722-4104

Star Gazer Instruments
Box 1851
Neepawa, MB ROJ 1HO
Canada
204-476-3855

Tectron Telescopes
2111 Whitfield Park Ave.
Sarasota, FL 34243
813-758-9890

Tele Vue Optics, Inc.
100 Route 59
Suffern, NY 10901
914-357-9522

University Optics
P.O. Box 1205
Ann Arbor, MI 48106
800-521-2828

Miscellaneous:

American Science & Surplus, Inc.
601 Linden Place
Evanston, IL 60202
708-982-0870

C & H Sales Co.
P.O. Box 5356
Pasadena, CA 91117-9988
818-681-4925

Copper & Brass Sales
6555 E. Davison
Detroit, MI 48212
313-365-7700

H & R Co.
18 Canal St. P.O. Box 122
Bristol, PA 19007-0122
800-848-8001

Hastings Irrigation Pipe Co.
P.O. Box 728
Hastings, NE 68902
402-463-6633

McMaster-Carr Supply Co.
Elmhurst, Illinois
708-834-9600

Small Parts, Inc.
13980 N.W. 58 Ct.
P.O. Box 4650
Miami Lakes, FL 33014-0650
305-557-8222

AUTHOR CREDITS

Kenneth Wilson contributed Chapters 1 through 8.

Robert Miller oversaw the projects, contributing Projects 1, 2, 6 and collaborating on 3.

Allan Saaf collaborated on Project 3 and contributed Projects 4 and 5.

Ronald Ravneberg contributed the additional project referred to as "Alice".

INDEX

Metric Equivalency

Inches	CM
1/8	0.3
1/4	0.6
3/8	1.0
1/2	1.3
5/8	1.6
3/4	1.9
7/8	2.2
1	2.5
1-1/4	3.2
1-1/2	3.8
1-3/4	4.4
2	5.1
2-1/2	6.4
3	7.6
3-1/2	8.9
4	10.2
4-1/2	11.4
5	12.7
6	15.2
7	17.8
8	20.3
9	22.9
10	25.4
11	27.9
12	30.5
13	33.0
14	35.6
15	38.1
16	40.6
17	43.2
18	45.7
19	48.3
20	50.8